Charles Johnson Maynard

Manual of taxidermy

A complete guide in collecting and preserving birds and mammals

Charles Johnson Maynard

Manual of taxidermy
A complete guide in collecting and preserving birds and mammals

ISBN/EAN: 9783741174698

Manufactured in Europe, USA, Canada, Australia, Japa

Cover: Foto ©berggeist007 / pixelio.de

Manufactured and distributed by brebook publishing software
(www.brebook.com)

Charles Johnson Maynard

Manual of taxidermy

MANUAL OF TAXIDERMY

A Complete Guide

IN COLLECTING AND PRESERVING
BIRDS AND MAMMALS

BY C. J. MAYNARD

ILLUSTRATED

SECOND EDITION.

BOSTON
S. E. CASSINO AND COMPANY
1884

ELECTROTYPED.

BOSTON STEREOTYPE FOUNDRY,
No. 4 PEARL STREET.

INTRODUCTION.

Twenty-five or thirty years ago amateur collectors of birds were rare ; in fact, excepting in the immediate vicinity of large cities, individuals who spent their leisure time in gathering birds for the sole purpose of study, were so seldom met with that, when one did occur, his occupation was so unusual as to excite the comments of his neighbors, and he became famous for miles around as highly eccentric. Such a man was regarded as harmless, but as just a little "cracked," and the lower classes gazed at him with open-mouthed wonder as he pursued his avocations ; while the more educated of his fellows regarded him with a kind of placid contempt. I am speaking now of the days when the ornithology of America was, so to speak, in obscurity ; for the brilliant meteor-light of the Wilsonian and Audubonian period had passed, and the great public quickly forgot that the birds and their ways had ever been first in the minds of any one. To be sure, men like Cassin, Lawrence, Baird, and Bryant were constantly writing of birds, but they did it in a quiet, scientific way, which did not reach the general public. Possibly the political troubles in which our country was involved had something to do with the great ornithological depression which fell upon the popular mind. Strange as it may appear, however, for a period of thirty years after the completion of Audubon's great work, not a general popular work of any kind was written on birds in America. Then appeared Samuels' "Birds of New England," published in 1867, a work which apparently did much toward turning the popular tide in favor of ornithological study, for from that time we can perceive a general awakening. Not only did the newspapers and magazines teem with articles on birds, but in the five succeeding years we find three important works on American ornithology announced as about to appear: Baird, Brewer, and Ridgeway's "History of American Birds," of which three volumes have appeared, published in 1874; Maynard's "Birds of Florida," issued in parts, but afterwards merged into the "Birds of Eastern North America," completed in 1882, and Coues' "Key," published in 1872. Other works quickly followed,

for now the popular ornithological tide was setting
strongly towards the flood, and it has ever since
been rushing on and gathering recruits as it goes,
until the tidal wave of popular favor for ornithological pursuits has reached from shore to shore
across our great continent ; and where there were
once only a few solitary devotees to this grand
science, we can number thousands, and still they
come ; so that high-water mark is not yet reached,
while to all appearances this tidal-wave will agitate
the coming generation more strongly than it does
the present.

Of all the vast numbers interested in the study
of bird life, there are few who do not gather specimens. Years ago, in the beginning of the study,
when the solitary naturalist had no one to sympa-
thize with him in his pursuits, birds' skins were
usually made in what we would now consider a
shocking manner. Within the last fifteen years,
however, since ornithologists have become more
numerous, and the opportunities of comparison of
workmanship in preserving specimens has been
facilitated, great improvements are seen. Slovenly
prepared collections are now far from desirable ; in
fact, even rare specimens lose much of their value
when poorly made up. When there are enough

experienced collectors in one locality to compare notes as to the various improvements each has made in skin-making and mounting birds, one aids the other; but there is always a multitude of be-ginners who live in isolated localities and who do not number experienced collectors among their friends, and who consequently require the aid of written instructions. Hence the need of books to teach them.

This little work, then, is intended to meet the wants of amateur ornithological collectors, wher-ever it may find them, for it is written by one who has at least had the advantage of a very wide ex-perience in collecting skins, making and mounting. He has also had the advantage of comparing his methods with those of many excellent amateurs and professional collectors throughout the country; and if he has not conferred any benefits on them, he has at least gained much useful information, and the results of all this are now laid before the reader.

The art of taxidermy is very ancient, and doubt less had its origin among the very early races of man, who not only removed the skins of birds and mammals for clothing, but also for ornaments. Birds and mammals were also frequently regarded as objects of worship, and consequently preserved after death, as among the ancient Egyptians, who embalmed entire birds and mammals that were considered sacred.

From the rude methods of preserving skins, doubtless, arose the idea of mounting, or placing the skins in lifelike attitudes. The first objects selected for this purpose were, of course, birds and mammals of singular forms or brilliant colors, as objects of curiosity. Later specimens would have been preserved for ornamental purposes, but it is probable that it was not until the seventeenth century that either birds or mammals were collected with any idea of their scientific value.

Specimens either mounted or in skins must have been rudely preserved at first, but, like all other branches of art and science, when people began to understand the value of well-made specimens when compared with those poorly done, workmen who became skilled in their art appeared and turned out good work. The art of making good skins, however, never was understood in this country, at least until within the last fifteen or twenty years, and even now it is rare to find good workmen who can make skins well and rapidly.

As is natural, many methods have been prac-
ticed to insure lifelike attitudes in birds and other
objects of natural history. A good opportunity of
studying the various schools of mounting may be
seen among the specimens of a large museum,
where material is gathered from various localities
throughout the world. I have seen birds filled
with many varieties of material, from cotton to
plaster, and have even seen cases where the skin
is drawn over a block of wood carved to imitate
the body removed.

As a rule, I prefer the soft body filling, where all
the wires are fastened together in the centre of the
inside of the skin, and cotton, or some similar elas-
tic material, filled in around it. This method is,
however, very difficult to learn, and, unless one has
had a large experience in handling birds, will not
give satisfactory results. I have therefore recommended the hard body method, as given in the
text, as being the best, as it is more easily learned
and always gives the best results in the hands of
amateurs.

In skin-making, although I have given two
methods, making in the form and wrapping, I prefer the latter, as being by far the best, although it
is not as easy to learn.

Mounting mammals and reptiles and making their skins also varies as given by different individuals, but I have given the method by which I have found, by experience, amateurs succeed the best.

Some may consider the information given in the following pages, too meagre for practical purposes, but I have purposely avoided giving lengthy instructions, considering a few well-worded sentences much better, as expressing much more clearly the ideas I wish to convey. In short, the reader has the condensed results of my extended experience, and if he will follow with care and patience the instructions herein given, I am sure that he will obtain satisfactory results from his labor.

I have endeavored to inculcate the idea in the following pages that he who wishes to be a successful taxidermist cannot accomplish his end without the utmost care; he must exercise patience and perseverance to the extreme; difficulties will arise, but he must overcome them by severe application to the study of his art, and, as years pass by, experience will teach him much that he never knew before. I have been assured many times, by men who are now skilful workmen, that their first ideas of preserving specimens were divined from my "Naturalist's Guide." Thus I trust the present little work may aid others who are entering the fairy land of science, to prepare lasting mementoes gathered by the way.

C. J. MAYNARD.

BOSTON, MASS.

TABLE OF CONTENTS.

LIST AND EXPLANATION OF PLATES.

FIG. 5. — PAGE 37.

Skull of bird, under side : Dotted lines A, A, A, show cuts to be made in removing a triangular piece of bone and muscle, to which the whole or a portion of the brain will adhere.

FIG. 6. — PAGE 42.

Dissection of a song sparrow, showing male organs of reproduction : 1 and 2, lungs ; 3, 3, testicles. The four organs below these are the kidneys.

FIG. 7. — PAGE 43.

Dissection of a song sparrow, showing female organs of reproduction : 4, lungs ; 1, 1, small yellow glands, present in both sexes ; 2, ovaries ; 3, oviduct. These last four figures are merely diagrams, only sufficiently accurate in outline to convey an idea of the position of the parts indicated.

FIG. 8. — PAGE 50.

Tweezers for making skins, mounting, etc. : Several sizes are used, but as a rule the points should be longer than those given in the cut.

FIG. 9. — PAGE 51.

Drying forms fastened to a board, D, skin in the form. I now use these forms detached. See text. Also, see page 54 for a better method of making skins which I now practise.

FIG. 10. — PAGE 54.

Form of a skin of an oriole : I now use the long label given on page 58. A skin should not be made too full ; a dead bird laid on its back will convey an idea of the thickness of the body of a skin.

Fig. 11. — Page 64.

Straight-nosed pliers : Used for bending wires in mounting.

Fig. 12. — Page 64.

Cutting-pliers : Used for cutting wires in mounting.

Fig. 13. — Page 66.

Body of a bird : E, neck-wire, which should be as long as the neck and tongue in order to reach into the upper mandible. This wire should be wrapped in cotton. B, wire before clinching ; G, C, wire clinched ; F, tail wire bent in the form of a T at H, a leg wire going through tarsus along dotted line to D.

Fig. 14. — Page 67.

Roughly-drawn skeleton of a pinnated grouse, only sufficiently accurate to indicate the different bones : A, skull ; B, B, B, vertebræ ; furcula of neck and back, or wishing-bone ; D, forearm ; F, carpus, showing hollow in bone through which the wire is to be passed in wiring the wing; G, end of furcula ; H, tip of keel ; I, indentations in posterior border of stemma ; J, femur ; K, tarsus ; L, heel ; M, pelvis ; N. cocyx ; O, crest of keel ; P, side of keel ; X, wire used in mounting skeleton ; A, B, ribs.

Fig. 15. — Page 69.

Outline figure of grouse showing external parts : A, back ; B, rump ; C, upper tail coverts ; D, under tail coverts ; E, ventral region ; F, tibra ; G, tarsus ; H, breast ; I, side ; J, throat ; N, chin ; L, abdomen ; M, feet.

Fig. 16. — Page 73.

Outline drawing of a mounted bird : A, A, dotted line to indicate the relative position of the head and body, with the perch on which the bird stands; B, B, winding cotton to keep the feathers in position; C, C, indicating proper position of wings; D, tail feathers "plated." I do not now recommend this method. E, E, tail bearing wire; F, upright of gland; H, horizontal bar of stand; I, feet of bird on stand; S, leg-wire wrapped around bar after emerging from foot.

FIG. 17. — PAGE 92.

Lower portion of bolt used in mounting large mammals: A, movable nut on screw C; B, immovable flat washer.

MANUAL OF TAXIDERMY.

PART I.—BIRDS.

CHAPTER I.

COLLECTING.

SECTION I. : TRAPPING, ETC. — Several devices for securing birds for specimens may be successfully practised, one of the simplest of which is the box-trap, so familiar to every schoolboy. If this be baited with an ear of corn and placed in woods frequented by jays, when the ground is covered with snow, and a few kernels of corn scattered about, as an attraction, these usually wary birds will not fail to enter the trap. I have captured numbers in this way, in fact, the first bird which I ever skinned and mounted, was a blue jay, caught in a box-trap. I was only a small boy then, so I do not now remember what first suggested mounting the bird, but the inherent desire to preserve the specimen must have been fully as strong then as in later years, or I never could have brought myself to the point of killing a bird in cold blood.

In fact, putting the bird to death is the worst of trapping; and with me, unless I do it at once, during the first excitement of finding the bird entrapped, the deed is likely never to be done at all. Sparrows, snow-buntings, and in fact nearly all birds of this class may be caught in box-traps in winter. For these small birds, scatter chaff over the snow so thickly as to conceal it, then use

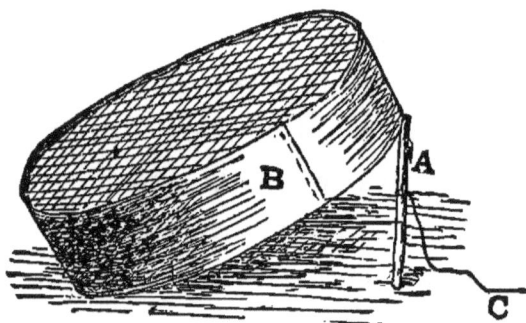

FIG. 1.

a spindle upon which canary-seed has been glued, for bait, scattering some of the seed outside. Other traps, however, may be used more success-fully for fringilline birds. For example, the clap-net trap, where two wings, covered with a net, close over the birds, which are attracted by seeds strewn in chaff, scattered in the snow. This trap, which is similar to those used by wild-pigeon catchers, is sprung by means of a long cord, the end of which is in the hands of a person who is concealed in a neighboring thicket or artificial bower. A very simple trap, but excellent for catching sparrows, may be made by tilting a common coal sieve on one edge, keeping it up by means of a stick which has a cord attached to the middle (see Fig. 1). The birds will readily go under the sieve, in search of food, when the trapper, who is concealed at a short distance, jerks out

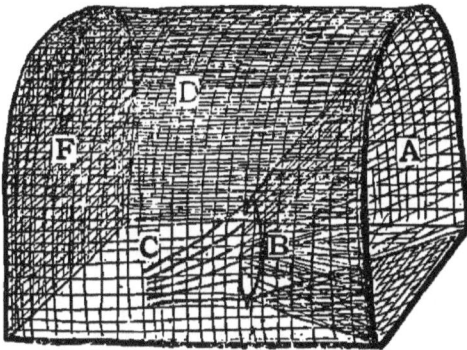

FIG. 2.

the stick by means of the cord ; the sieve falls and the birds are captured. This trap requires constant watching, which, in cold days, is not very pleasant ; thus a much better trap may be found in one of my own inventions, which is called the " Ever-ready Bird Trap." It is made of strong netting stretched over wire, and is placed on the ground or on a board in a tree. A decoy bird, of the same species as those to be captured, is procured if possible, and placed in the back of the trap at Fig. 2, and then the birds enter the front of the trap, B; pass through the way of wires, C, which pointing backward after the manner of the well-known rat-trap, prevent their egress. This trap is constantly set, and several birds are captured at one time. Orioles, bobolinks, rose-breasted grosbeaks, goldfinches, snow-buntings, all other sparrows and finches, in fact, all birds which will come to a decoy or bait, may be taken in this trap.

I have frequently taken jays in small snares similar to those used in capturing rabbits. Quail and ruffed grouse were also taken in this manner before the present time, but it is now illegal to trap game-birds in nearly all the States.

The steel trap of the smallest size is exceedingly useful in capturing hawks, owls, and even eagles, as well as many other large birds. One way is to set it in the nest of the bird, first taking care to remove the eggs, substituting for them those of a hen. Almost all large birds may be taken in this manner, and it is an excellent way to identify the eggs in case of some rare hawks or herons. The topmost portion of some dead stub, which is a favorite roost of a hawk or eagle, is a good place to set a trap; and small hawks and owls may be captured by putting the trap on the top of a stake, some eight or ten feet high, in a meadow, especially if there are no fences near. Hawks and owls haunt meadows in search of mice, and invariably light upon a solitary stake, if they can find one, in order to eat their prey or to rest, and thus are very apt to put their "foot into it," in a manner decidedly agreeable to the collector, if not so pleasing to themselves. Steel-traps may also be set on boards nailed to trees, in the woods or on hill-tops, but they should in this case be baited with a small mammal or bird. I have succeeded in capturing marsh hawks by tying a living mouse to a steel-trap, and placing it in a meadow which was frequented by these birds. Other hawks and also eagles may be captured by using decoys; the best thing for this purpose being, strangely enough, a live great horned owl. The owl is fastened to a stout stake in an open field or meadow during the migration of hawks, in the spring or fall, and surrounded by baited traps. The hawks passing over are attracted by the novel spectacle of an owl in such a peculiar position and come swooping down for a nearer view, when they perceive the bait, and in trying to eat it are caught. A hawk or eagle may be used in this way as a decoy, but the great horned owl is by far the best.

In using steel traps, care should be taken to wrap the jaws with cloth, so as to prevent injury to the legs of the bird captured. Vultures may be taken in steel traps by simply baiting them with any kind of flesh. Many species of birds may be successfully captured by one or another of the methods given. In fact, we are in constant receipt of trapped birds during the proper seasons, and thus many hawks and owls which would have been difficult to procure are taken in numbers by our collectors.

Bird-lime, although scarcely advisable when the birds are intended to be preserved, may be used to advantage in capturing birds for the cage. A small quantity of it is spread on a twig or small stick, one end of which is lightly stuck in a notch on some upright branch or stem, in such a position that the bird must alight on it in order to reach the bait. The stick should be poised so lightly that the slightest touch of the bird's feet will cause it to drop, when the bird, giving a downward stroke with its wings to save itself from falling, will strike the outer quills against the stick, and thus both feet and wings become fastened to it by the adhering lime. In case of a rare specimen, the lime may be removed from the plumage by the aid of alcohol, or the bird will remove it in time, if permitted to live. Good bird-lime is difficult to procure ; that made from linseed-oil and tar, boiled down, is the best ; but this process must be carried on in the open air, as the mixture is exceedingly inflammable. The sticky mass thus obtained must be worked with the hands under water, until it assumes the proper consistency. In spreading lime on the sticks, the fingers should be wet to prevent the lime sticking to them. Another way in which I have taken such unsuspicious birds as pine grosbeaks, crossbills and red-polls, is by placing a noose of fine wire on the end of a pole, and by approaching a tree cautiously, in which the birds were feeding, have managed to slip it over their heads, when they are drawn fluttering downward, and the noose removed, before any permanent injury is done. I have even taken pine grosbeaks in an open field in this manner, and have ascended a tree and captured them with only the noose attached to a stout piece of wire, in my hand.

SECTION II. : SHOOTING.—Although, as shown, many valuable species may be secured by trapping, snaring, etc., yet the collector relies mainly

on his gun. This much being decided, it at once
occurs to the beginner, What kind of a weapon
shall I get? Of course, muzzle-loaders are now
out of the question; and among the multitudes of
breech-loaders in the market, one has only to
consult his taste or the length of his purse.
Therefore it is simply useless for me to recommend any particular make of gun. Good single-
barrel breech-loaders can be bought for from nine
dollars to twenty dollars, while double-barrels cost
from fifteen dollars upward. For ordinary collecting, a twelve-gauge is perhaps better than any
other, as such birds as ducks, hawks and crows
can be readily killed with it. For warblers,
wrens, and other small birds, however, a much
smaller gauge gun is almost indispensable, as a
large gun sends the shot with such force that it
not only penetrates the body of the bird, but also
goes out on the opposite side; thus each shot.
makes two holes, when one is all that is necessary to kill. This fact should then always be
kept in mind, and as a rule load lightly, with just
enough powder to cause the shot to penetrate well
into the bird without going through it. In a
twelve-gauge gun, two drachms of powder behind
an ounce of shot is sufficient to kill a bird like a

jay or golden-winged woodpecker, at a distance of thirty or forty yards; then if more penetration is necessary, more powder may be used with the same quantity of shot, but this will cause the shot to scatter more. A good collecting gun, one which will kill small birds with a very small amount of ammunition and little noise, has long been a desideratum. I have tried many kinds, but nothing has proved so satisfactory as a small repeating gun of my own invention, and which is manufactured by us. This gun consists of two brass tubes, a smaller one within a larger, with an air space between, thus greatly deadening the sound; and both are securely fastened to a finely nickel-plated five-shot revolver. We make two sizes, a twenty-two gauge, the report of which is very slight, and a thirty-two gauge, which makes a little louder noise. The former will kill warblers at fifteen yards, and the latter at twenty yards, while birds like jays, thrushes, and robins, may be brought down with the thirty-two gauge at a distance of ten yards. This gun served me well in Florida last winter, and I killed at least two-thirds of the birds that I collected there with it. The light report of such a gun does not frighten the birds, while the fact that one nearly always

has a second shot ready in the revolving cylinder,
is a great help, in case of a wounded bird, or in
the sudden appearance of a second specimen, as
so often happens, after the first has fallen. The
price of this gun varies from four dollars and fifty
cents to five dollars and seventy-five cents, accord-
ing to quality and size. Blow-guns, air-guns,
catapults, etc., are useful only in cases when a
shot-gun cannot be used, as they cannot be depended upon. A collector, in order to procure
birds with a certainty, requires a good shot-gun.
The ammunition used in the small collecting gun
is copper shells, primed, of three lengths for each
size. For shot, I use dust numbers ten and eight,
but for a larger gun, coarser shot is sometimes
necessary ; collectors, however, — especially begin-
ners, — are apt to use too large shot. On the con-
trary, I do not like to shoot too fine shot at large
birds ; thus a hawk killed with a heavy charge of
dust-shot at twenty yards would have the feathers
cut up very badly, whereas a warbler shot at the
same distance would be likely to make a good
specimen, as it would only receive a few pellets of
shot, whereas a large number would strike the
hawk. As a rule, then, use dust-shot for birds up
to the size of a cedar-bird, then number ten to the

size of a jay, after which number eight will kill better and cleaner, and I should use this size as long as it will bring down the birds ; and it is surprising to see how large species may be killed with it. I have taken brown pelicans, wild geese, and large hawks with number eight, and I once secured a frigate-bird with it, all at good distances. For very large birds like cranes, white pelicans, or eagles I have used a rifle very successfully. A thirty-two gauge Allen is my favorite gun, and I have killed birds at all distances from twenty to three hundred and twenty-five yards with it. Of course, nearly all successful rifle shots must be made at sitting birds, as I have met with but few who could bring them down when flying. Another good method of securing large shy birds which go in flocks, is to load with buckshot, putting a stiff charge of powder, say three to five drachms, behind it, then fire into the flock from a distance, elevating the gun at an angle of some forty-five degrees above the birds. I have killed both species of pelicans at two hundred yards distant in this way.

SECTION III.: PROCURING BIRDS. — Birds are to be found nearly everywhere, in fact, there is scarcely a square acre of land on the face of the earth which is not inhabited, at one season or

another, by some species, and many are found on
the beaches, and on the ocean itself. Following
are some of the localities in which our American
species are to be found; and, presumably, foreign
birds of the same families will occur in similar
places.

TURDIDÆ : THRUSHES. — Of these, the robin is
the most common and is found everywhere. Next
among the true thrushes are the olive-backed,
hermit, and allied species. These occur usually
in woodlands, and are rather shy, keeping at a distance. The wood thrush inhabits deeply-wooded
glens. The mocking thrushes prefer thickets in
the neighborhood of dwellings, — for example, the
cat-bird. The brown thrush also inhabits thickets, but are not, as a rule, fond of the society of
man, while the smaller thrushes, of which the
golden-crowned is an example, prefer the woodlands; and the two water thrushes are found in
swampy localities.

SAXICOLIDÆ : STONE-CHATS. — The blue-birds
are often sociable, building in orchards and farm-
yards, while the western species appear to prefer
mountain cliffs as breeding-places. The rare stone-
chat is, I think, found in open sections where it
occurs at all.

CINCLIDÆ: OUZEL. — The solitary species of ouzel found with us inhabits the mountain streams of the far west.

SYLVIDÆ : TRUE WARBLERS. — Are pre-eminently birds of the woodlands, but occasionally the kinglets, notably the golden-crowned, will wander into orchards during mild days in winter.

CHAMÆIDÆ : WRENTITS. — The only species found in the United States inhabits the sagebrush in the far southwest.

PARIDÆ : TITMICE. — Are also found in the woods or thickets, but some species wander into the orchards during winter.

SITTIDÆ : NUTHATCHES. — Are birds of the woodlands as a rule, but the white and red-bellied nuthatches wander considerably in autumn, while the brown-headed seldom if ever leave the piny woods of the south.

TROGLODYTIDÆ : WRENS. — The creeper-wrens are found among the cacti of the far southwest, while the rock-wrens occur among thickets in a similar region. The true wrens are found in thickets, often in the neighborhood of dwellings, in which they frequently build, while the two marsh wrens occur on both salt and fresh water marshes throughout the country.

ALAUDIDÆ: TRUE LARKS. — These birds occur on the far prairies, on the coast of Labrador, and in winter along the barren seashores of the northern and middle section.

MOTACILIDÆ : WAGTAILS. — Are also birds of the open country, and the titlark is found in fields during the migrations, especially along the coast from Maine to Florida.

SYLVICOLIDÆ : AMERICAN WARBLERS. — These gems of the woodland and of wayside thickets abound throughout the length and breadth of our country. During the migrations they are generally distributed, it not being uncommon, then, to find even the Blackburnian warbler, which, during the nesting season, is pre-eminently a bird of the deep woods, feeding in the open fields, while I have taken the Cape May warbler, which occurs in summer in the thick evergreens of the north, feeding among the oranges and bananas in the gardens of Key West. Warblers then should be looked after nearly everywhere, among willows by the brookside, on the barren hill-tops which scarcely support a scant growth of pine or cedars, and on the blooming trees of orchards. Some species are exceedingly shy, so as to require a heavy charge of dust-shot to reach them, while others are so tame as to peer inquisitively into the very face of a collector as he makes his way through their chosen retreats.

TANAGRIDÆ: TANAGERS. — These strikingly colored birds are usually found in the woods, occasionally however visiting the open sections. They are rather shy and retiring in habits, and their presence must be usually detected by their song.

HIRUNDINIDÆ: SWALLOWS. — Are birds of the open country, and are more common in the vicinity of settlements than elsewhere. The violet-green swallow, however, occurs among the cliffs of the Rocky Mountains.

AMPELIDÆ: WAXWINGS. — Are, as a rule, found in the open country in the vicinity of settlements; and even the Bohemian waxwings occur abundantly in some of the cities of Utah in winter, feeding upon the fruit of the ornamental trees.

VIREONIDÆ: VIREOS. — These widely-distributed birds are usually fond of the woodlands, but the white-eye prefers thickets in swampy places, while the warbling is seldom found far from settlements; indeed, more often inhabits trees which grow in the streets of villages than other sections.

LANIIDÆ: SHRIKES. — Are found in open sections, often in fields, and on the uninhabited Indian

hunting-grounds of Florida. I found the logger-
heads along the borders of the open prairies.

FRINGILLIDÆ : FINCHES, SPARROWS, AND GROS-
BEAKS. — These are, as a rule, found mainly in
the more open country. The cross-bills, however, enter thick woods, especially evergreens.
The grosbeaks, notably the rose-breast, prefer the
woodlands. The blue sparrows, like the indigo
bird, are found in open fields grown up to bushes.
The snow-buntings occur in open fields and along
barren sections of seaboard, while the sharp-tailed
and seaside finches inhabit the marshes. The
grass sparrows, notably the yellow-winged, Hens-
low's, and Leconte's, prefer grassy plains. Last
winter I procured all three species of this genus
(*Coturniculus*) on a plantation in Western Florida,
securing them all in three successive shots, a feat
which has, I am certain, never before been accomplished. Many of these grass-haunting birds. have
to be shot as they rise from the herbage to fly
away, but I found, by persistingly following a.
specimen from point to point, that after a time
it would settle in a bush, when I could secure it
with my repeating collecting gun.

ICTERIDÆ : ORIOLES, BLACKBIRDS, ETC.—Orioles
prefer, as a rule, orchards and ornamental trees

about dwellings, but they sometimes occur in the more open woodlands. The marsh blackbirds, like the red-wings and yellow-headed, prefer wet meadows. The rusty and brewer's are found in swamps. The crow blackbirds and boat-tailed occur in fields and along the borders of streams.

CORVIDÆ : CROWS, JAYS, ETC. — These usually occur in the woodlands or thickets. Crows frequent the seashore in numbers in winter, and may be secured by exposing meat which is poisoned by strychnine, as they will frequently eat it during the inclement season. Canada and blue jays occur in woods, while the Florida and California jays inhabit thickets.

TYRANNIDÆ : FLYCATCHERS. — Are widely distributed species. The king-birds are found in the . more open sections, and the same is true of the crested flycatchers. The bridge pewee inhabits the vicinity of dwellings, while the wood pewee occurs in the woods. The least flycatcher prefers orchards, but the greater portions of the genus *Empidonax* are found in woodlands or thickets.

CAPRIMULGIDÆ : GOATSUCKERS. — The whip-poor-wills and chuck-wills-widow occur in the thick wood, emerging occasionally at night, but seldom straying from their retreats. A good way to secure these birds is to note as accurately as possible the point where one begins to sing; then, on the following evening, conceal yourself near the spot, when the bird will be seen to emerge from its retreat and alight on some particular rock, post, or branch, on which it invariably perches, and utters its song. Then if the bird be too far away to secure at the time, it may readily be taken another evening by the collector posting himself nearer. These birds may also be started from their concealment during daylight, and thus be shot. The night-hawks inhabit the more open sections, but perch on trees during the day. They may readily be secured while flying over the fields.

CYPSELIDÆ: SWIFTS.—The white-throated swift occurs among the clefts of the Rocky Mountains, and is exceedingly difficult to procure. The well-known chimney swift inhabits chimneys almost everywhere, but, as it never alights outside of these retreats, must be shot on the wing.

TROCHILIDÆ : HUMMINGBIRDS. — Inhabit as a rule the open country. I have secured numbers of our ruby-throats on cherry-trees when they were in blossom, and later, on beds of flowers ; and I presume the western species may be found in similar situations. I shoot them with light charges of dust-shot, fired from my collecting gun.

ALCIDINIDÆ: KINGFISHERS. — These noisy birds are found plentifully in the vicinity of streams. They are shy and require a heavy charge of number eight to bring them down.

CUCULIDÆ: CUCKOOS. — The roadrunner of California, Texas, and intermediate locality, occurs in the sage bush, but our species of cuckoos, even the mangrove, inhabit thickets from which they occasionally emerge. They are usually betrayed by their notes. They are easily killed, their skin being very thin and tender.

PICIDÆ: WOODPECKERS. — Occur, as a rule, in the woodlands, but the smaller species and the golden-winged inhabit orchards. They are all tough birds to kill. They are a generally distributed family, but some species are confined to certain localities, for example, the great ivorybilled is not found outside of Florida, and even there, is confined to a limited area, and very rare. Strickland's woodpecker has as yet only been found in the United States in a single range of mountains in Arizona.

PSITTACIDÆ: PARROTS.—Our Carolina paroquet is now exceedingly rare out of Florida, and then occurs in the neighborhood of cypress swamps, but occasionally visits the plantations.

STRIGIDÆ : OWLS. — The burrowing owl occurs in the western plains and in a limited area of Florida. The snowy owl inhabits sand-hills of the coast in winter, and the short-eared occurs in the marshes, but all other species are birds of the deep woods, occasionally emerging, however, especially at night. The great horned and barred may be decoyed within shooting distance in the spring by imitating their cries, and the latter-named species will also eagerly fly toward the collector when he produces a squeaking sound similar to that made by a mouse. The small owls may be often found in holes of trees.

FALCONIDÆ : HAWKS, EAGLES, ETC. — Marsh hawks occur in fields, meadows, and marshes. Everglade kites are found on the widespread savannahs of Florida, while the swallow-tailed Mississippi and white-shouldered are found on the prairies of the south and west. The buzzard hawks usually occur in the woods, but during the migrations pass over the fields, flying high. The fish-hawk is abundant on the seacoast, but also visits the ponds and lakes of the interior. The duck-hawk is fond of clefts, and migrates along the seacoast. The sharp-shinned sparrow and pigeon are often found in solitary trees in fields, where they hunt for mice, but they also occur in open woods. The bald eagle occurs on the seashore or on large bodies of water, but the golden eagle prefers the mountainous regions.

CATHARTIDÆ : VULTURES.—Occur everywhere throughout the south. The great California vulture is now very rare.

COLUMBIDÆ : PIGEONS.—Are usually found in fields, but the wild pigeon is often taken in the woods. The ground doves are found in fields which are bordered with thickets, to which they retreat when alarmed. Two or three species are found on the Florida Keys, and about as many more in Texas.

MELEAGRIDÆ : TURKEYS.—Wild turkeys occur in the wilderness of the south and west. They inhabit open woods as a rule, often roosting at night in swamps.

TETRAONIDÆ: GROUSE, QUAIL, ETC.—The Canada, ruffled, and allied species of grouse occur in the woodlands. The prairie sharp-tail and sagehen are found on the plains of the west, while the ptarmigans inhabit the bleak regions of the north. The common quail is widely distributed through·

out the more open country, from Massachusetts to
Texas, and the plumed California and allied species
occur in the southwest, frequenting the thickets of
the prairies, or along the mountain-sides.

CHARADRIIDÆ: PLOVERS. — These are, as a rule,
maritime birds, especially during the southward
migrations, but many of the species breed in the
interior, and the kildeer and mountain plovers are
always more common on bodies of fresh water.
None of the species are, however, found far from
water, but they all alight in dry fields in search
of food.

HÆMATOPODIDÆ: OYSTER-CATCHERS AND TURN-
STONES. — All these birds inhabit the seacoast.
They occur in oyster-beds or among rocks.

RECURVIROSTRIDÆ: AVOCETS AND STILTS. —
Both these species are birds of the interior, being
found in the south and west in the vicinity of
water.

PHALEROPODIDÆ: PHALEROPES. — These singular birds are found off the coast, often far out at
sea during winter, but, oddly enough, breed in the
interior, nesting throughout the northwest and
north. They are, however, occasionally found on
the coast during the northward migration, especially
during storms.

SCOLOPACIDÆ: SNIPES, WOODCOCK, ETC. — Woodcock and snipes are usually found in fresh-water swamps, especially in spring. The true sandpipers, like peep, grass-birds, etc., haunt the pools in marshes or accompany the sanderlings on the beaches. The godwits are found on the marshes, as are also red-breasted snipe, but the curlews inhabit hill-tops, especially during the autumnal migration. I have, however, found the long-billed curlew on the beaches of Florida. Willets and yellow-legs occur on the marshes or on the borders of streams.

TANTALIDÆ: IBISES AND SPOONBILLS. — Occur along the borders of streams and other bodies of fresh water, or on mud-flats in the far south.

ARDEIDÆ: HERONS. — These are widely dis-tributed birds. The true herons occur along the margins of bodies of water, both on the coast and in the interior, while the bitterns generally haunt only the fresh water.

GRUIDÆ: CRANES. — Are found on the prairies of the west and south, frequenting the vicinity of water.

ARAMIDÆ: COURLAN. — The well-known crying-bird is found only in Florida, inhabiting swamps along the rivers and lakes of the interior.

RALLIDÆ: RAILS, GALLINULES, AND COOTS.— The true rails inhabit very wet marshes, both salt and fresh, concealing themselves in the grass. Gallinules and coots are found on the borders of fresh water.

PHŒNICOPTERIDÆ: FLAMINGOES.—The flamingo occurs only with us, on the extensive mud-flats in extreme Southern Florida, where they are exceedingly difficult to procure, being very shy.

ANATIDÆ : GEESE, DUCKS, ETC. — These are all inhabitants of the water, being seldom found far from it. Some species, like the teal, prefer secluded pools in the interior, while the wood-duck and others frequent woodland streams ; and the eiders and marine ducks are abundant in the waters of the ocean.

SULIDÆ : GANNETS. — Excepting while breeding, these birds keep well out to sea, and are thus quite difficult to procure. All of the marine species are liable to be driven inland during severe storms, and the collector should not fail to take advantage of such circumstances.

PELICANIDÆ: PELICANS. — The brown pelican is a resident of the extreme southern coast, and may be found on sand-bars or perched on trees in the immediate vicinity of water. The white pelican is found in similar localities in winter, but migrates northward during the summer, breeding in the interior, from Utah to the Arctic regions.

GRACULIDÆ: CORMORANTS. — Occur on sand-bars in the south, or on rocky cliffs in the north, and on the Pacific coast. During migrations they keep well out to sea. They have the habit, in common with the gannets and pelicans, of alighting on barren sand-spits which rise out of the water.

PLOTIDÆ: DARTERS. — The snake-bird of the south occurs on bodies of fresh water, and may be seen perched on trees or flying high in air. They are exceedingly difficult to kill, being, as a rule, shy, and very tenacious of life.

TACHYPETIDÆ: FRIGATE BIRDS. — The frigate bird is found with us only on the Gulf of Mexico and among the Florida Keys. They are usually seen upon wing, but I have observed thousands perched on the mangroves on the Keys. They roost on the trees on lonely islets at night, at which time they appear so stupid that they may be approached quite readily.

PHÆTONTIDÆ: TROPIC BIRDS. — These fine birds occur only in tropical waters unless they are accidentally blown out of their latitude by

storms. They breed on the rocky cliffs of the
Bahamas and Bermudas.

LARIDÆ : GULLS, TERNS, ETC. — The Skua gulls
keep well out to sea, as a rule, but occasionally
enter harbors and bays in pursuit of gulls and
terns, which they rob of their prey. Gulls and
terns of the various species rest on sand-bars or
fly along the shore.

PROCELLARIDÆ : PETRELS. — Excepting while
breeding, these birds keep well out to sea and are
thus quite difficult to procure. They haunt the
waters which are frequented by fishermen, however, and may be procured by visiting these localities on some fishing-smack.

COLYMBIDÆ : LOONS. — Are found in both fresh
and salt waters, but are somewhat difficult to procure on account of their habit of diving.

PODICIPIDÆ : GREBES. — These birds have similar habits to those of the loons, but are found in
smaller bodies of water, notably the Pied-billed,
one or more specimens of which occur in almost
every little pool throughout the country, especially
during the southward migration.

ALCIDÆ : AUKS, PUFFINS, ETC. — These birds
are found off the coast during migration, but breed
on the rocky shores of both coasts.

Although the foregoing list gives the locality in which a given species may be found, as a rule, it is always well to bear in mind that birds have wings, and by the use of them may stray into unaccustomed localities far distant from their usual habitance. For example, a burrowing owl was shot on the marshes of Newburyport, and a petrel, which has hitherto been known to science through a single specimen which was taken many years ago in the southern hemisphere, was picked up, in an exhausted condition, in a ploughed field of the interior of New York. The young collector then should ever be on the alert, keeping well in mind the fact that the art which he is pursuing is not lightly learned. I have frequently heard the inexperienced remark that he could easily kill a hundred birds in a day; and although this might be true on certain occasions, — for I have seen over this number killed by one person in two discharges of a gun, — yet, as a rule, a good collector will seldom bring in over fifty birds during his best days. A man must not only be experienced, but will be obliged to work hard in order to average twenty-five birds in a day. Although there are some " born " collectors who will procure birds, even if they be provided with no

more formidable weapon than a boy's catapult,
yet the peculiar attributes which make up a good
collector are mainly to be acquired. A quick
eye to detect a flutter of a wing or the flit of a
tail among waving foliage; an ear ready to catch
the slighest chirp heard amid the rustling leaves,
and so skilled as to intrepret the simple grada-
tions of sound which distinguish the different
species; a constant wide-awake alertness, so that
nothing escapes the observation, and which gives
such nice control over the muscles that the gun
comes to the shoulder with a promptitude that
combines thought with action; and an unwearying
patience and pluck which totally disregard minor
obstacles, are some of the characteristics which
must be possessed by the individual who wishes
to bring together a good collection of birds by
his own exertions. If one does not possess these
traits, why, then study to acquire them; for
securing birds is as fine an art as is preserving
them after they are obtained.

SECTION IV. : CARE OF SPECIMENS. — Just as
soon as a bird is shot, examine it carefully by blow-
ing aside the feathers in order to find the shot-
holes; if they bleed, remove the clotted blood with
a small stick, or, better, the point of a penknife,

then with a pointed stick, or the knife, plug the hole with a little cotton, and sprinkle plaster, or better, some of my preservative, on the spot. Next plug the mouth with cotton, taking care to push the wad down far enough to allow the bill to close, for if the mandibles are left open the skin of the chin and upper throat will dry, causing the feathers to stand upright. Smooth the specimen lightly and place it, head down, in a paper cone, which should be long enough to allow folding the top without bending the tail feathers. Then the bird may be placed in a fish basket, which is the best receptacle for carrying birds, as it is not only light to carry, but also admits the air. Never shut a bird up in a close box in warm weather, as it will spoil very quickly. Care of a bird in the field will save much labor, and your cabinet specimens will look enough better to warrant it. Blood left under the plumage gradually soaks through the feathers, thus causing them to become matted, when they are exceedingly difficult to clean. Some specimens however, will bleed, and if they are to be preserved this blood must be removed. I have always found it best to wash the blood off in the first water I could find, and then let the bird dry, either by carrying

in my hand, or, by suspending it to a limb of
a tree, where I could return for it afterwards.
Care should be taken in such cases, however, to
wash *all* the blood off, and then plug the wound
with cotton, as if any flows out when the plumage
is wet it will spread on the feathers and stain
them. In picking up birds that are only wounded
never take them by the tail, wing, or any part of
the plumage, but grasp them firmly in the hand
in such a way as to imprison both wings, then
kill them by a firm pressure of the thumb and
forefinger, applied to the sides just back of the
wings. This compresses the lungs, and the birds
die of suffocation almost instantly. Never strike
a bird, no matter how large, with a stick, but in
case of hawks, eagles, etc, the talons of which are
dangerous, seize them first by the tip of one
wing, then by the other, work the hands downward until the back is grasped, then apply the
pressure to the lungs. There is no danger from
the beak of even the most formidable species
after the pressure is put upon the lungs, for I
never knew a specimen to bite while being killed
in this way; the only thing necessary is to
keep out of the way of their talons. I have frequently been obliged to remove eagles from a box

and kill them, and have done it with my hands
alone.

Wounded doves and pigeons should be grasped
very firmly, and not allowed to struggle in the
least, as their feathers fall out very easily; and
the same is true, though to a less extent, with
cuckoos; in fact, it is always best to brush the
plumage as little as possible, handling the specimen when dead by the feet or bill. In picking up
white herons or other birds which have fallen in
mud or other dirty water, take them up by the bill
and shake them gently to remove the ooze. The
feathers of all birds, especially aquatic species,
are covered with a delicate oil, and all extraneous
matter glides off the plumage if they are not
soaked in water. In catching wounded herons,
take them by the beak to avoid the danger of
losing an eye from a lunge of the sharp point.
When a bird is to be placed in a basket or on a
bench, do not *throw* it down, but lay it gently on
its back, always bearing in mind that the smoother
a bird is kept before it is skinned the better it will
look when preserved. I have even noticed that
the true ornithological enthusiast always keeps
his birds in good condition, while others who
merely shoot birds for the momentary pleasure of

the thing, or for gain, are very apt to handle them roughly. In other words, the student of nature possesses an innate love of his pursuits, which causes him to respect even a dead bird.

CHAPTER II.

SKINNING BIRDS.

SECTION I.: ORDINARY METHOD. — The only instruments that I use in removing the skin of birds ordinarily is a simple knife of a peculiar form (see Fig. 3) ; but I like to have a pair of dissecting scissors by me to be used in cases given further

FIG. 3.

on. I also have plenty of cotton, and either Indian meal or dermal preservative at hand to absorb blood and other juices.

To remove the skin from the bird, first see that the mouth is plugged with cotton, and if it is, note if this be dry, if not remove it and substitute fresh. It is also well to note if the bird be flexible, for if rigid it is extremely difficult to skin, and it is always best to wait until this peculiar rigidity of the muscles, which follows death in all vertebrate animals, shall have passed. This occurs in warm weather in much less time than in

33

cold, often in one or two hours, but in moderate temperature a bird had better lie for at least six hours after it has been killed.   Take then a specimen in the proper condition, lay it on its back on a bench, on which clean paper has been spread, with its head from you, but slightly inclined to the left.   Now part the feathers of the abdomen with the left hand, and, excepting in ducks and a few other species, a space, either naked or covered with down, will be seen extending from the lower or costal extremity of the sternum to the vent. Insert the point of the knife, which is held in the left hand, with the back downward, under the skin near the sternum, and, by sliding it downward, make an incision quite to the vent, taking care not to cut through the walls of the abdomen. This can readily be avoided in fresh birds, but not in specimens that have been softened by lying too long.   The fingers of the right hand should be employed during this operation in holding apart the feathers.   Now sprinkle meal or preservative in the incision, especially if blood or juices flow out, in order to absorb them and prevent them soiling the feathers.   Next, with the thumb and finger of the right hand, peel down the skin on the left side of the orifice, at the same time pressing the tibia

on that side upward. This will disclose the
second joint of the leg, or knee proper. Pass the
knife under this joint, and, by cutting against the
thumb, cut it completely off, a matter easily accomplished in small birds ; rub a little absorbent
on either side of the severed joint ; then grasping
the end of the tibia firmly between the thumb and
forefinger of the right hand, draw it outward. At
the same time, the skin of the leg should be
pressed downward by the fingers of the right hand
to prevent tearing. The leg is thus easily exposed,
and should be, as a rule, skinned to the tarsal
joint. With the thumb-nail, nip off the extreme
tip of the tibial bone, and strip the flesh off the
remainder of the bone by a downward pull ; then
give the whole a twist, and cut all the tendrils at
once. Of course the flesh may be removed from
the bone by scraping, etc., but the above is the
best method, and in case of large birds, break the
end of the tibia with pliers. Turn the bird end for
end, and proceed the same with the other leg, but
during both operations the bird should not be
raised from the bench. Now peel away the skin
about the tail, place the forefinger under its base,
and cut downward through the caudal vertebra
and muscles of the back quite to the skin, the

finger being a guide to prevent going through
this.   Rub absorbent on the severed portion.
Grasp the end of the vertebra protruding from
the body, thus raising the bird from the bench;
peel down front and back by pushing downward
with the hand, rather coaxing the skin off than
forcing or pulling it.   Soon the wings will appear;
sever these where the humerus joins the cora-
coid, cutting through the muscles from above
downward in large specimens, thus more readily
finding the joints.   Rub on absorbent, and it may
be well to remark that this must be done when-
ever a fresh cut is made.   Then the body is laid
on the bench, and the skin is held in one hand, or,
in large specimens, allowed to rest on the lap or
on the bench, but never to dangle.   Keep on
peeling over the neck by using the tips of as many
fingers as can be brought into service and soon,
the skull will appear.   The next obstruction will
be the ears; these should be pulled or, better,
pinched out with the thumb and forefinger nails.
Do not tear the ears, and special care should
be exercised in this respect in owls.   When
the eyes are exposed, pass the knife between
the lids and orbit, close to the former, taking
care that the nyctatating membrane be removed

from the skin, or it will be in the way when
the eyelids are arranged in making the skin.
Peel well down to the base of the bill, so that
every portion of the skin may be covered with
preservative. Push the point of the knife under
the eyes, and remove them by a single motion,

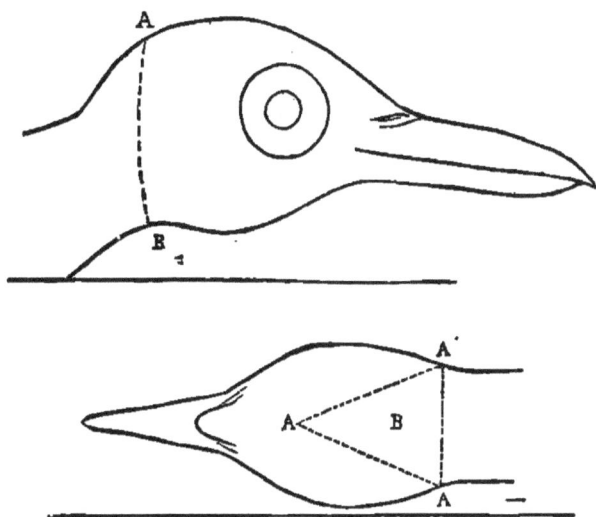

FIG. 5.

without breaking them. Cut off the back of the
skull at the point shown in the line A, Fig. 4;
Turn the head over and make two cuts outward as
seen at A. A., Fig. 5. thus removing a triangular
portion of the skull B, Fig. 4, to which the brain
will usually adhere, but when it does not, remove
it with the point of the knife. This leaves the
eye-cavities open from beneath. Draw out the

wings by grasping the end of the humerus in
the left hand, and press the skin back with the
right, to the forearm ; then with the thumb-nail,
or back of the knife, separate the secondary quills
which adhere to the larger bone from it, thus
turning out the wing to the last joint or phalanges.
Cover the skin well with preservative, especially
the skull, wings, and base of tail ; roll up balls
of cotton of about the size of the entire eye
removed, and place in the cavities in such a condition that the smooth side of the ball may come
outward so that the eyelids may be arranged
neatly over them.   Nothing now remains but to
turn the skin back to its former position.   Turn
the wings by gently pulling the primaries and the
head, by forcing the skull upward until the bill
can be grasped; then by pulling forward on this,
and working the skin backward with one hand,
the matter will be accomplished, when the feathers
may be lightly smoothed and arranged.   It must
be borne in mind that the quicker and more
lightly a skin is removed the better the specimen
will look.   By lightly, I mean that the skin should
not be tightly grasped nor stretched by pulling.
Some workmen will remove a skin from a bird
which is nearly spoiled without starting a feather,

while others may skin a specimen as quickly, but the plumage will be crushed and broken through rough usage. The time for removing the skin from a small bird should not exceed six minutes, and I have seen it taken off in half this time. Of course the beginner will be longer than this; and then the skin should be occasionally moistened, by using a damp sponge.

SECTION II.: EXCEPTIONS TO THE USUAL METHOD OF SKINNING. — In case of birds which are very soft on account of having been dead a long time, it may be advisable to open either beneath the wing, making a short incision along the side or above the wing, cutting along the feather tracks just above the scapularies; and some skin ducks through a hole in the back just above the rump. I do not, however, advise such practice, as a rule, as the skins are more difficult to make up, and the bird cannot be mounted quite as readily.

Woodpeckers with large heads and small necks, like the pileated and ivory-billed, and ducks having similar characteristics, as the wood, pintail, and a few other species; also flamingoes, sand-hill, and whooping cranes, cannot be skinned over the head in the usual manner, but the neck should be cut off after the skin has been removed as far as pos-

sible, and then a slit should be cut in the back of
the head, and the head be skinned through this
orifice, but an abundance of absorbent should be
used to prevent the feathers from becoming soiled.

Care should be exercised in skinning cuckoos,
doves, thrushes, and some species of sparrows, as
the skin is not only thin, but the feathers start in
the rump and back very readily. Peel the skin off
gently, and do not fold it abruptly backward in
working on these parts, but hold it as nearly as
possible in its original position. The skin of the
wood duck, and sometimes that of the hooded
merganser, adheres to the flesh of the breast, but
it may be separated by working carefully with the
back of the knife. In removing the skins of young
birds in the down, like ducks and gallinaceous birds,
do not attempt to skin the wings.

If a specimen is to be mounted with the wings
spread, the secondaries should not be detached,
but the knife should be forced down back of the
primaries in order to break up the muscles ; then
as much of the flesh as possible should be removed,
and a quantity of preservative pushed in beneath
the skin. In larger birds a slit should be made
on the under side of the wing, and the muscles
removed from the outside without detaching the

secondaries; and also when a specimen is to be
mounted, the eye cavities should be filled with clay
well kneaded to the consistency of putty.

SECTION III.: ASCERTAINING THE SEX OF BIRDS.
— Although the sex of many birds can be ascertained with tolerable certainty by the plumage,
yet this is never an infallible guide, and to make
perfectly sure of every case the internal organs
should be examined. I always advise dissecting
such plainly-marked birds as scarlet tanagers or
red-winged blackbirds, and by practising this habit
I was once fortunate enough to discover a female
painted bunting in full male livery. The sex of
birds can be readily ascertained in the following
manner : Lay the bird's body on its left side, with
the head from you ; then with a knife or scissors,
cut through the ribs and abdominal walls on the
right side ; then raise the intestines, and the organs
will appear.

In males, two bodies, the testicles, more or less
spherical, will be seen lying just below the lungs on
the upper portion of the kidneys (Fig. 6, 3, 3). These
vary not only in color from white to black, but also
in size, depending upon the season or age of the
the specimen. Thus, in an adult song sparrow,
during the beginning of the breeding season, the

testicles will be nearly or quite a half inch in diameter, whereas in autumn they will not exceed a
number eight shot in size; and in nestlings of the
same species they are not larger than a small pellet
of dust-shot. At this early age, the sex of birds
which have become somewhat soft is quite difficult
to determine, and the same is true at any season

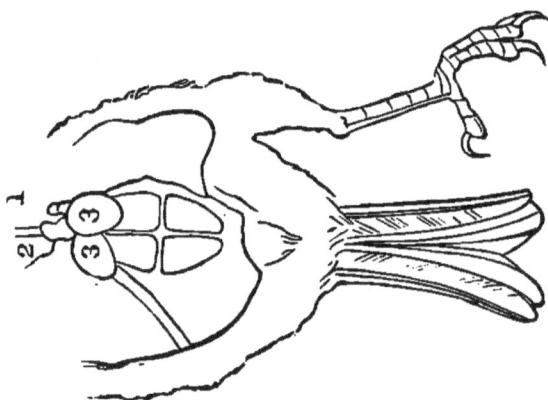

FIG. 6.

if the specimens be badly shot up. There are
other organs, however, in the male. For example,
the sperm ducts are always present, appearing like
two white lines; and in the breeding season the
plexus of nerves and arteries about the vent becomes swollen, forming two prominent tubercles
on either side (Fig. 6, 3, 3).

In the female the ovaries lie on the right side
(Fig. 7, 2) in about the same position as is occupied

by the testicles in the male. The ovaries vary in
size from that of half the size of an egg to minute
points, depending, as in the male, on the season of
the year and age of the specimen. In very young
birds the ovaries consist of a small white body
which under a magnifying glass appears somewhat

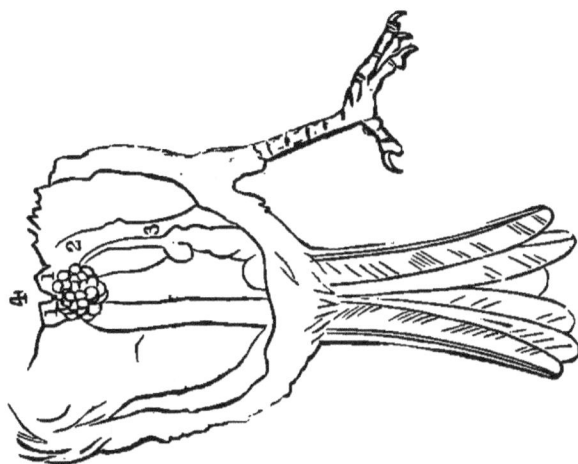

FIG. 7.

granular. In both male and female are two yellow-
ish or whitish bodies, in the former sex lying above
the testicles, but further forward, and consequently
just in front of the kidneys; and in the female they
occupy about the same position. In addition to
the ovaries in the female, the oviduct is always
present (Fig. 7, 3), large, swollen, and convoluted
during the breeding season, but smaller and nearly

straight at other times. In young specimens it appears as a small white line.

The denuded breast and abdomen seen in birds during the breeding season, cannot always be depended upon as a mark of sex, as this occasionally occurs in males as well as in females.

SECTION IV. : PRESERVING SKINS. — Táxidermists for many years have made use of arsenic in some form as a preservative ; and in the first edition of my "Naturalists' Guide," I recommended the use of it dry, stating that I did not think it injurious if not actually eaten. I have, however, since had abundant cause to change my opinion in this respect, and now pronounce it a dangerous poison. Not one person in fifty can handle the requisite quantity of arsenic necessary to preserve specimens, for any length of time, without feeling the effects of it. For a long time I was poisoned by it, but attributed it to the noxious gases arising from birds that had been kept too long. It is possible that the poison from arsenic with which my system was filled might have been affected by these gases, causing it to develop itself, but I do not think that the gas itself is especially injurious, as I have never been poisoned since I discontinued the use of arsenic.

When I became convinced that arsenic was

injuring my health, and that of others, I began to
experiment upon other substances, and after trying
a quantity of various things, have succeeded in
manufacturing a nearly odorless compound which
has the following advantages over arsenic: It
thoroughly preserves the skins of birds, mammals,
reptiles, and fishes from decay, and also prevents
the attacks of dermestes or anthrenus, while the
feathers of birds and hair of mammals are not as
liable to be attacked by moths as when the skin is
preserved with arsenic. This preservative when
properly applied abstracts the oil from greasy skins,
thus preventing them from becoming decayed
through carbonization, as nearly always occurs in
ducks' skins after a few years. It is a deodorizer,
all disagreeable smells leaving the skin to which it
is applied ; and above all it is not a poison. I used
this dermal preservative, as we have named it, as
an absorbent while skinning birds, especially small
ones, as then the plumage is dusted with it neces-
sarily, which insures more or less protection to the
feathers from the attacks of moths.

To render my preservative, or indeed any other,
effective, it must be thoroughly applied to the
skin; all the portions, especially those to which
any flesh adheres, must be well covered with it,

and the fibre of the muscles should be broken up
as much as possible. But a small portion, at best,
of arsenic is soluble in either water or alochol, and
but a little in the juices of the skin, whereas in
my dermal preservative at least three-fourths of
that which comes in contact with a moist skin is
absorbed, thus thoroughly preserving the specimen. In the case of a greasy skin, remove as
much fat as possible by peeling it off or gently
scraping until all the little cells which contain the
oil are broken up and the skin appears ; then coat
the skin liberally with the preservative, when it
will be found to absorb the oil. Allow this layer
to remain a few minutes, then scrape it all off and
coat again with a fresh supply. Continue to do
this until all the oil that will flow out is absorbed,
and then dust with a final coating.

There are two chemical processes carried on
in preserving oily skins, one of which converts the
oil into soap, and this is in turn absorbed and
dried. Thus the preservative which has been
scraped from the skin can be after a time used
again, as it has lost but a small portion of its effi-
ciency. It might be borne in mind, however, that
all the fat cells possible must be broken up, as
the skin which surrounds these is, in a measure,

impervious to the preservative, which must, in order to absorb oil, come in contact with it.

SECTION V. : OTHER METHODS OF PRESERVING SKINS— Skins may be temporarily preserved by simply using black pepper, but the effect is not lasting. The same is true of tannic acid, but either of these, alum, or even common salt, will do as a substitute for the preservative until the skins can be got into the hands of a taxidermist, or until the collector can procure the proper preservative. I will here mention that the dermal preservative costs only twenty-five cents per single pound, and this quantity will preserve at least three times as many skins as the same amount of arsenic.

A good method by which large skins may be temporarily preserved is by salting them. Simply coat the inside of the skin with fine salt, turn it, smooth the feathers and fold the wings neatly, then pack in paper. The salt prevents the skin from quite drying, and thus it can be moistened much more readily, and made into a skin or mounted. The advantage of packing large birds in so small a capacity is obvious to any one. Two collectors whom we have had out the past season have sent in some thousand large skins in this condition ; and these we shall endeavor to work up

within six months' time, as salted skins become
quite brittle if allowed to lie too long. They
should be kept in a dry place, as salt absorbs
moisture, which causes the skin to decay. They
are also liable after the first year to be attacked
by dermestes and anthrenus.

Birds which are in a bad condition through
having been dead a long time may be sometimes
skinned, in case of rare specimens, by using great
care. Sprinkle the inside of the skin well with
preservative, as this tends to set the feathers,
being a stringent, keeping the skin as straight as
possible, as folding it is liable to loosen the
feathers. The intestines of birds may be removed
and the cavity salted when large birds are to be
sent from a distance.

CHAPTER III.

SECTION I.: CLEANING FEATHERS. — If a bird is bloody, the feathers may be washed either in turpentine or water. Saturate a rag or piece of cotton, and clean off the blood, which if dry may require some soaking. Try to keep the water from spreading as much as possible, but be sure that every particle of clotted blood is removed and the spot washed thoroughly. Then dry by covering the spot well with either plaster or dermal preservative, the latter being preferable as it never bleaches the plumage. This should be worked well into the feathers with a soft brush, aided by the fingers, applying a fresh supply constantly until all the moisture is absorbed; then dust with a soft duster. In case of grease-spots, if fresh, use the dermal preservative alone, but if old and yellow use benzine to start the grease, and then dry with preservative, when it will generally be found that all stains will be removed; but in some cases two or three applications of benzine may be neces-

49

sary. Small spots of dried blood may often be removed from dark feathers by simply scraping with
the thumb-nail, aided by a moderately stiff brush,
much after the manner in which a living bird
removes foreign substances from its plumage. Do
not leave clotted spots of blood in the plumage, as
the feathers never lie well over them, and such
places are liable to be attacked by insects, and
even a spot of blood under the wing should, in
my opinion, always be removed. Before any at-

FIG. 8.

tempt is made either to make a bird into a skin
or mount it, it should be thoroughly cleansed.
Stains of dirt may be removed with alcohol, which
dries more readily than water, but it will not start
blood as well as turpentine or water.

SECTION II.: MAKING SKINS OF SMALL BIRDS.—
The instruments for skin-making are a flat brush,
a duster for cleansing, three or four pairs of tweezers of varying sizes (see Fig. 8), needles, curved
or straight as preferred, silk thread for sewing,
and soft cotton for winding, and metal forms made
of rolled tin or zinc (Fig. 9). Lay the skin on its

back, and push the single bones left on the forearm into the skin, then fasten them by taking a
stitch through the skin near the base of the wing ;
then, passing the thread around the bone, tie it
firmly. Now with the same thread, uncut, sew the
other bone in a similar manner, leaving the two
connected by a piece of thread which is about as
long as the natural width of the body of the bird,
thus the wings are kept the same distance apart as

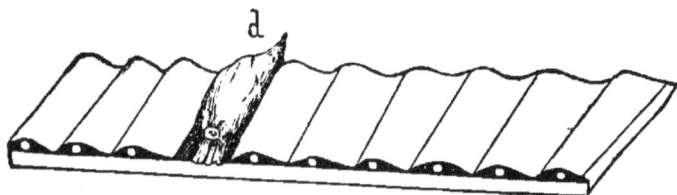

FIG. 9.

they were formerly. Now take a piece of cotton
and form it into a rough body as near as possible
in size to the one removed, but having a tapering
neck of about the length of nature. Now grasp
this firmly in the tweezers, and place it, neck fore-
most, in the skin, taking care that the point of
the tweezers enters the brain cavity of the skull,
so that the cotton may fill it, and projecting downward, form the throat; now allow the tweezers
to open, and slip them out. Open the eyelids,
arranging them neatly over the rounded cotton

beneath. See that the bones of the wing lie
along the sides, as they are liable to become
pushed forward in putting in the cotton. This
can be remedied by raising the cotton gently. If
the cotton body has been placed in the proper
position the neck will be full, but not over stuffed,
and of just the right length to form a skin that
has the appearance and size of a freshly-killed bird
lying on its back with the head straight. The bill
should be horizontal with the bench on which the
bird lies, and from which the specimen should not
be raised while at work on it. Now roll the skin
over and examine the back ; see that the wing
feathers, especially the scapularies, lie in regular
rotation, and that they have not been pushed one
above the other ; and the same attention should be
given to the tail. Note if the feathers of the back
lie neatly over the scapularies, and these in turn,
should be over the wing-coverts ; in short, all
should blend neatly, forming a smoothly rounded
back. Now place the skin, back down, in the
form, lifting, by placing the thumb and forefinger
on either side of the shoulders, which is the
proper way to handle a small skin, even when dry.
In placing the skin in the form, care should be
used that the cotton does not slip out of the skull,

causing the head to fall down. See if the tip of the wings are of equal length; if not make them so by drawing one wing downward, and pushing the other up toward the head, but do not pull them out of place at the shoulders. Be careful that the wings are placed high enough on the back. This is easily ascertained, if the closed tips of the primaries lie perfectly flat on the bottom of the form with their inner edges nearly downward. Now smooth the feathers with a pair of tweezers, placing the feathers of the sides that come below the sparrow's wing inside the wing; above this they will lie outside. Always bear in mind that although a skin can be made perfectly smooth by an expert in from eight to fifteen minutes, one who is not accustomed to the work will be obliged to occupy a much longer time, as a skin cannot be made too smooth. Arrange all spots and lines on the feathers as they occur in life, especially about the head or on the back; in fact, too much attention cannot be given to these details, before and after a skin is placed in the form, if one wishes to turn out a first-class specimen.

Now bind the skin with soft cotton thread, used on bobbins in the mills, beginning at the lower portions of the wings, and winding the thread over

the body and under the form, so that the threads
lie about a quarter of an inch apart, ending with
the throat. Now arrange all the feathers which
may have become disarranged under the threads,
and place the skin away to dry where there is no
draft, for a slight breeze will be sure to blow some
of the feathers out of place. (For the form of a
skin, see Fig. 10.)

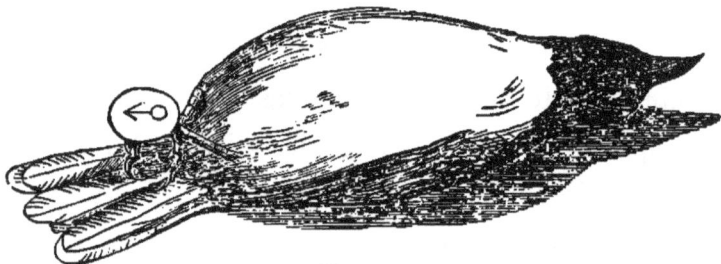

FIG. 10.

Another method of making skins which may be
practised to advantage is as follows : After the
skin is ready to place in the form, wrap it closely
in a *very* thin layer of nice cotton batting, taking
care that the feathers lie perfectly smooth, although
these may be partially arranged through the cotton, which must be thin enough for the feathers
to be seen through it. The skin is then laid aside
to dry without placing in the form.

Skins should not be exposed to too great artificial
heat, neither should they be left to dry during

damp weather in a room without a fire. Small birds, like warblers, will set perfectly hard in forty-eight hours in a moderate temperature with dry air. Never allow a skin to freeze.

SECTION III.: MAKING SKINS OF LONG-NECKED BIRDS. — Sandpipers, thin-necked woodpeckers, or any birds, the necks of which are liable to become broken, should have a wire placed in the neck to support and strengthen it. Proceed in sewing the wing-bones as directed in small skins; then make a body of cotton around the end of a wire that has about an inch of the end bent into the form of a hook, and then the body may be wrapped about the wire with some of the winding cotton. The neck-wire should project from the body for about the same length as the natural neck, or a little more. This neck-wire should also be wrapped with cotton to the size of the natural neck, but rather thicker where it joins the body. A small portion of this wire which has been sharpened, as hereafter to be directed, should project beyond the body. Now place the body in position inside of the skin, forcing the point of the wire into the skull, up into the base of the upper mandible as far as it will go. The heads of long-billed birds may be turned on one side, but in this case the bill will be placed

to a greater or less angle. Sew up specimen
as before; arrange and place in a long form and
bind. The legs of such birds as yellow-legs may
be stitched together at the tibial joint, then bent
toward the sides, and the toes stitched to the skin.
In making skins of all birds where the back of
the head is opened, the orifice should not be sewed
up until after the wire has been inserted in the
upper mandible, as it may be necessary to add
more cotton through here to make the throat
or back of the head as full as in life. Sew up this
orifice by taking fine stitches in only the extreme
edge of the skin, and the same caution must be
exercised in sewing up accidental tears in the
skin. Very tender skins may have tears mended
by pasting tissue-paper neatly over the holes from
the inside. In fact it is best to sew up tears
from the inside, always using silk thread for the
purpose.

SECTION IV.: MAKING SKINS OF HERONS, IBISES,
ETC. — Proceed exactly as in long-necked birds,
but to make a compact skin lay the bird breast
down, and turn the head and neck on the back,
and fasten the legs to the sides. I always wire
the necks, and for additional security, to prevent
them being straightened by careless or inex-

perienced persons, I stitch the bill to the skin of
the back. In addition to sewing on the inside of
the wing, stitch the wing firmly to the inside, by
sewing over the outer primary into a pinch of skin
on the side, thus the wing is fastened in two places.

Ducks' skins may be made in a similar manner,
but the feathers of the side must be brought *over*
the wings, and the webs of the feet may be spread
with a wire, which must be removed, however,
when the feet are dry, or it will rust ; and galvan-
ized or brass wire is the best for making skins.

SECTION V. : HAWKS, OWLS, EAGLES, VUL-
TURES, ETC. — The skins of these large birds are
made in forms, but the wings must be stitched to
the sides, as in herons, etc. The necks must be
wired. In making the skins of all large birds it
is best to use bodies made of excelsior or grass,
rather than cotton, which does not make a firm
enough body. See remarks under mounting for
instructions for making bodies ; but they do not
need to be quite as solid for skins as in mounting ;
in fact, keep them as light as possible. Too much
care cannot be taken in forming the eyelids of all
birds, especially large ones. Have the cavity occupied by the eye round, with the cotton lying
smoothly inside, and not projecting in a ragged
manner.

SECTION VI. : LABELLING SPECIMENS. — A skin is of little value unless labelled with date, locality, and sex. Never lay a bird one side without a label is firmly attached to one foot or other part. The sex of birds is indicated by the astronomical signs of the planets ; Mars (♂) and Venus (♀), the former being, as is obvious, the mark for males and the latter for females. To keep these in mind one has only to remember, that that of Mars is a conventionalized spear and shield, indications of his warlike profession, while that of Venus is supposed to represent a looking-glass, an article so indispensable to feminine taste. I use blank forms for labels, and the simpler the better ; thus, below is one which I used during my last expedition to Florida : —

> EXPLORATIONS IN FLORIDA,
> By C. J. Maynard & Co.,
> 9 Pemberton Square, Boston, Mass.
> *Rosewood, Nov. 10, 1881.*   ♂

The sex of either, male or female, is printed, but at least two-thirds as many males as females are needed ; while any notes regarding the color of feet, bill, and iris of each specimen may be written on the back. The size given is the one used for specimens from the size of a humming-bird to that of a golden-winged woodpecker. The labels of ducks and herons may be attached to the beak by securing through the nostrils, as then they are more readily found.

It is well to keep in mind that in order to have any value as a scientific specimen, a bird must be labelled as near as possible with date, locality, and sex, but never guess at either. If you have a skin in your possession that you are not absolutely certain about, either label it with an interrogation mark filling the part of which you are in doubt, or do not label it at all. Thus if you are unable to determine the sex satisfactorily, say so by draw-ing a line through the sex mark and substituting a query (?).

SECTION VII.: CARE OF SKINS, CABINETS, ETC.— When skins are removed from the forms they should be dusted with a light feather-duster, striking them gently from the head downward so as not to ruffle the plumage. Although skins are well preserved from the attacks of demestes and anthrenus, which feed upon the skin, yet the feathers are always liable to be attacked by moths, while the skin on the feet or bills is also liable to be eaten. This may be prevented by washing the

parts with a solution of bleached shelac dissolved
in alcohol. By far the best way to insure absolute
safety is to shut up the skins in insect-proof
cabinets. Various methods have been tried to
prevent the ingress of moths, etc., in cabinets,
but the best and simplest is to have a door
fitted to the outside of the drawers of an other-
wise perfectly jointed cabinet. This door is
provided with a bead which surrounds the outside and fits in a groove on the margin of the
woodwork outside the draws, while the whole door
fits in a groove which extends quite across the
bottom. Another method which we practise on
our latest-made cabinets is to have each drawer
moth-proof, by having a margin made all around it
which fits into a groove, then all the drawers are
covered by closing a flange on the sides.

SECTION VIII. : MEASURING SPECIMENS.—Specimens of all rare birds should be measured. With
the beginner, it is best to measure every specimen.
I measured some fifteen thousand birds before I
made a single skin without so doing, and now I am
careful to take the dimensions of all rare specimens. The dimensions of a bird are taken as
follows, using dividers and a rule marked in hundredths of the inch : First measure the extreme

length from the tip of bill to end of tail; then the
extreme stretch óf wing from tip to tip; then the
length of one wing from the scapular joint to tip
of longest quill; next, the length of tail from
end of longest feather to its base at the insertion
in the muscles; now the length of bill along
culmen or chord of upper mandibles; and of
tarsus from tarsal joint to base of toes. I have
a blank sheet ruled, and fill it out as per sample
(page 62).

SECTION IX.: MAKING OVER OLD SKINS. —
Sometimes it is desirable in case of rare birds
to make over into presentable skins specimens
which have been improperly prepared. Prepare a
dampening box by placing a quantity of sand,
dampened so as to just drip water, in any metal
vessel having a tight-fitting cover. Then wrap
the specimen to be made over in paper, lay it on
the sand, and cover with a damp cloth folded
several times. Place the cover on the vessel and
set in a moderately warm place for about twenty-
four hours if the specimen be small, longer if
large. At the end of this time the skin will be
quite pliable. Then remove the cotton and
examine the inside of the skin carefully, and if
there are any hard places caused by the skin

CAMPEPHILUS PRINCIPALIS.

No.	Sex.	Locality.	Date.	Length.	Stretch.	Wing.	Tail.	Bill.	Tarsus.	Color of			Remarks.
										Eye.	Bill.	Feet.	
1936	♂	Gulf Hummock, Fla.	Nov. 20, 1882	20.35	31.00	9.30	6.35	2.75	1.80	Yellow	Ivory white	Greenish	Plumage, new
1937	♀	"	"	19.75	30.00	9.00	6.25	2.65	1.60	"	"	"	"
1938	♂	"	"	21.00	32.00	9.60	6.50	2.80	2.00	"	"	"	"

being too thick, scrape them down with a blunt knife or, better, use our skin-rasp, and thus thin them down until the feathers above are as flexible as in any other portion. If there be grease on the feathers or inside of the skin after scraping, wash with benzine, and dry with preservative as described. When every portion of the specimen is perfectly pliable, and all superfluous dried flesh has been removed, sew up the rents, and make up as in fresh birds, but such skins generally require more careful binding. It is also often necessary to wire the neck of even small birds, especially in badly shattered and decayed skins.

CHAPTER IV.

MOUNTING BIRDS.

Section I.: Instruments. — The instruménts necessary for mounting are cutting pliers (Fig. 12), or tin shears, straight-nosed pliers (Fig. 11), wire

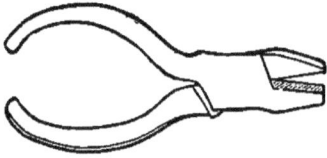

FIG. 11.

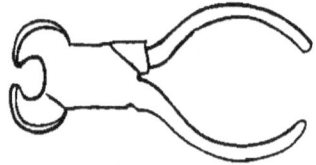

FIG. 12.

of various sizes, tweezers, and other implements used in skin-making; leg awls, for dried skins, and awls for boring stands; also stands of various kinds.

Section II.: Mounting from Fresh Speci-mens. — Be sure that a skin is perfectly clean in every way before attempting to mount, as it cannot be washed nearly as well afterwards. Remove all the bodies of skinned specimens well out of the way, and spread a clean sheet of paper where the skinning has been done, that there may be no danger of soiling the plumage. Make a body of fine grass, excelsior, or, better, the peculiar

tough grass which grows in shady places, in sandy
soil, is preferable, by winding with thread, moulding
it so as to have it quite solid, shaping it in the
hands until it assumes the exact length and
breadth of the body removed, and as near its form
as possible. Thus see that the back is fuller
than the under side, and that there is a well-
defined breast. Great care should be taken not to
get this body larger than the natural one; if
anything it should be smaller. With the pliers
cut a piece of wire of the proper size, that is, of
about half the diameter of the bird's tarsus, and
about three times the length of the body. In
cutting all wires which are to be sharpened, the
cut should be made diagonally across it, thus
forming a point. Push this wire through the
body so that it will emerge in the front much
nearer the back than the breast, protruding so
that it equals the length of the neck and tongue
of body removed. Bend over the end remaining at the back, turn down about half of it and
force it into the body (Fig. 13, c). This will hold
firmly if the body has been made sufficiently solid.
Wrap the wire with cotton by taking a strip and
winding it gradually so that it assumes a tapering form with a portion of the wire protruding.

Place this body in the skin and push the protruding wire into the upper mandible. Cut two
wires of about half the size of that already used,
and twice the length of the outstretched wing.
Work these into the wings, beginning at the
fleshy portion of the phalanges, so on into the
body, taking care not to allow it to pierce through

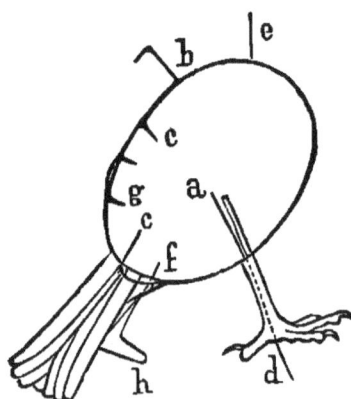

FIG. 13.

the skin anywhere. The wire should enter the body
at the point where the end of the lower portion
of the forearm touches it when the wing is folded
naturally. Pass the wire through the body diagonally until it emerges so that it can be grasped
with the pliers somewhere near the orifice, and
firmly clenched. Next find the metacarpal bone,
which has a hollow place in the centre (Fig. 14, f),
and force the upper end of the wire through it so

that about a quarter of inch shall protrude on the
upper side of the wing, and bend this down by
applying one jaw of the flat pliers on the side of

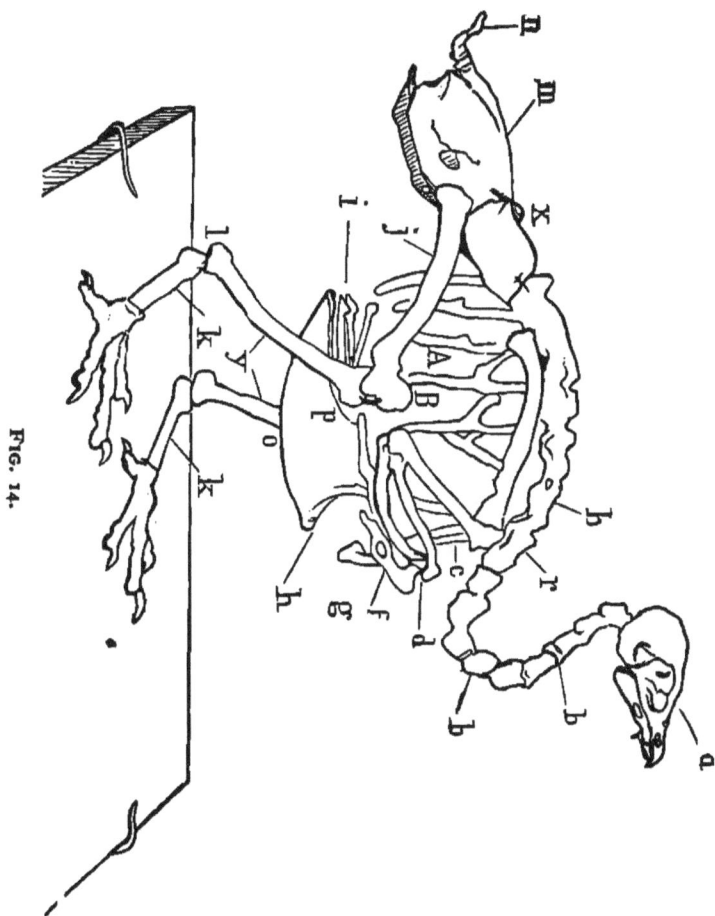

FIG. 14.

the wing opposite. This will fasten the wing
firmly, and the spurious wing will cover the wire,
while that on the lower side will be concealed

by the feathers. The wing should be outstretched
when this is done.

Cut wire for the legs of the same size as used for
the neck, and about as long. Pass them up through
the tarsus, inserting in the middle of the sole of
the foot. Be sure the wire is perfectly straight
before attempting this. A good way to straighten
wire is to place a pine-board on the floor, stand on
it, and then draw a long pull of wire under it by
grasping the end with pliers ; or a small piece of
wire may be straightened by rolling it on the
bench with a file. If the skin of the tarsus splits
in boring, it shows that the wire used is either
too large or crooked. After the wire is pushed up
to the heel or tarsal joint (Fig. 15, f), turn the
tibial bone out until the point of the wire appears,
when it should be grasped and drawn up so that
the point protrudes slightly beyond the tibial
joint. Wrap the tibial bone, wire and all, with
cotton or tow (in large specimens, the wire should
be bound to the bone with fine wire or thread) so
as to form a natural leg, then draw it back into
the skin. Next force the wire through the body
at the point where the knee touches it, or about
midway on the side. The wire will emerge on the
opposite side. Turn down the skin of the orifice,

draw the wire out, leaving about enough projecting out of the sole of the foot to go through the
perch of a stand and clench ; then fasten the end
firmly into the body. On large birds, like eagles,
I draw the wire through the body twice before

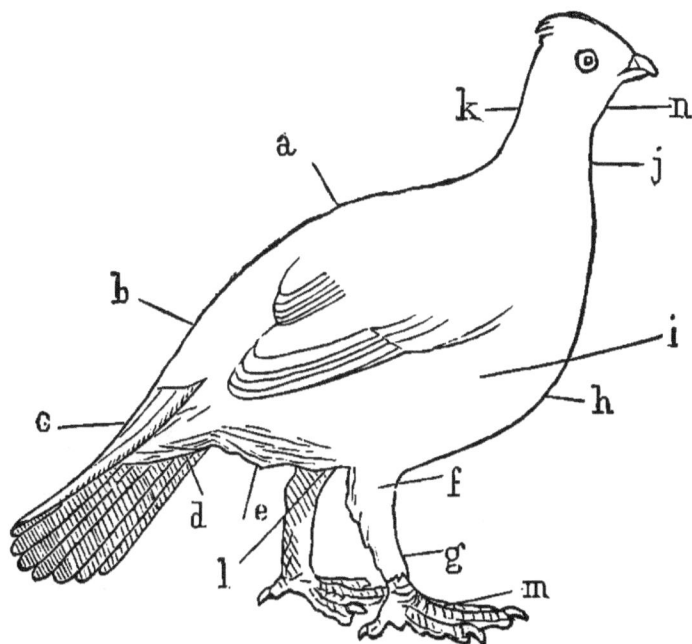

FIG. 15.

clenching, to make all secure. This work must
be well done if the bird is to be mounted nicely, as
it must stand firmly on its feet. As a rule, use
wire large enough, at least, to support the weight
of the body and skin without bending, but wire
one-half the size of the tarsus is generally large

enough to do this.   Cut a tail-wire which is at
least as long as the entire bird.   Insert it under
the tail, so that it enters the muscles in which the
feathers are embodied, taking care that it does not
spread them apart; push this up the centre of the
body so that it will emerge at an angle just at the
upper portion of the orifice, and clench it.   Bend
the remaining end under the tail twice, so as to
form a T, on which the tail may rest, and which
should, however, have the top broad enough to
spread the tail on to the required width.   During
wiring see that the plumage is ruffled as little as
possible ; also avoid soiling by keeping the specimen on clean paper.   If by chance the feathers
become greasy, they may be cleaned by sprinkling
liberally with the dermal preservative, which is
afterward brushed off.

Sew up the orifice neatly, taking care, as before
described, only to take in the extreme outer edge
of the skin ; and, if the body be not too large, it
will meet nicely.   If the body has not been made
quite large enough, especially on the breast, some
cotton may be placed between the skin and body
before sewing.   This must be done neatly, with
tweezers however, not so as to form a wad, but
spread out so as to blend neatly with the curve of the

body. Now place the wires which protrude from
the feet in holes bored in the perch of the stand,
which should be about as far apart as the bird naturally stands while perching. See that the feet
come well down on the perch with the toes arranged
properly, remembering that cuckoos, woodpeckers,
etc., have two toes in front and two behind, while
with hawks, owls, etc., the outer toe generally stands
at right angles with the others, and should therefore
grasp the end of the stand. Either twist the ends
of the wire together or wind them around the
stand very firmly. Now comes the most difficult
part of the task of mounting. Hitherto all has
been merely mechanical; certain rules had to be
observed only. But now the instructor must
pause for want of words wherewith to express his
meaning, for who can tell an artist how to put in
those bold and hasty strokes with which he maps
out his picture? He knows just what he is about,
however, for he has before his mental vision the
complete picture, and strives to place on canvas
that which appears before him. So must the artistic
taxidermist have before him a vision of the bird he
wishes to represent, with the combined mass of
feathers now in hand. Whether lightly poised for
flight or calmly sitting at rest, before he puts his

hand to the work before him let him fully decide what he wishes to produce. Let him see it just as clearly as he sees the birds sporting in their natural element. The true artist does not copy what the imagination of others have produced, he invents for himself or takes nature as his guide. Let us then who aspire to the highest in taxidermal art, take infallible nature as our guide. Study carefully every poise of the birds, every uplifting of the wing, every turn of the head or motion of the eyelids. I have long made a practice of keeping birds in confinement in order to thoroughly impress on my mind the different attitudes which they assume. I have had nearly all species of our owls, hawks, and eagles, and have kept herons, gulls, terns, pelicans, auks, and almost countless numbers of smaller birds, and in this way I have become so familiar with them that I can tell at a glance whether a bird is mounted in an easy attitude. Well, there must be no hesitation in mounting birds, or the specimens will dry ; and I will merely state in what order I arrange the different members, then leave the attitudes to my pupils. I first see that the bird stands correctly, that the legs are bent so that the bird will balance well in the position in which I

wish it to be placed. As a rule, a perpendicular
line drawn through the back of the head of a
perching bird will fall through its feet (see Fig.

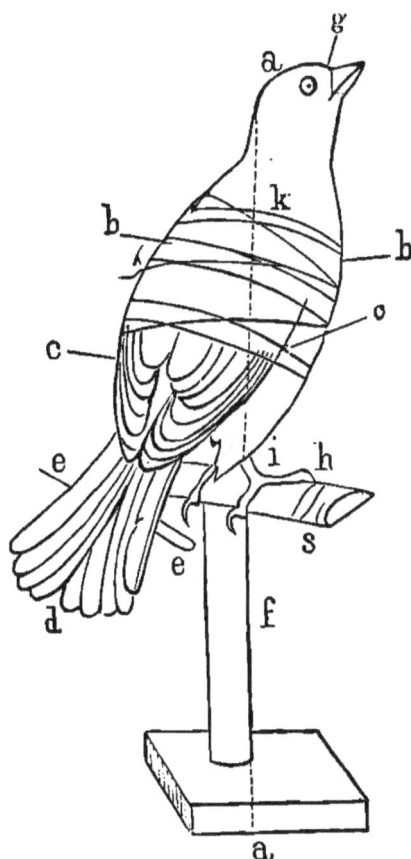

FIG. 16.

16, *a a*). Now bring the bird into position, and
fold the wings just as the bird does it. Note if the
scapularies, tertiaries, and secondaries lie in their

proper places, the first highest and the others
under them, which will give the bird a good
rounded back. Now place the bird in the proper
attitude, with the neck properly bent, remembering that in nearly all birds this nearly assumes the
form of the letter S, especially in long-necked
species. I do not like to see a bird staring
straight forward, but, as this is a mere matter of
fancy, I will not presume to dictate regarding
attitudes, only make the specimen look easy. Be
artistic, even if the specimen is going into a
public museum, where birds too often stare at the
visitors in grotesque attitudes. One can be
interesting and easy even in writing on the driest
scientific subject, — why not then give ease and
grace to our museum specimens? No more room
need be occupied ; a slight turn of the head, a
twist of the neck, or an advance of a foot, will do
this just as a bird would do it if it were alive.
Now place the eyes in position, and these should be
pushed well into the clay, and the lids arranged
over them naturally with a needle. Do not have
the eyes too large, as it gives the bird a staring ex-
pression, nor too small, but as near as possible to
the natural ones removed. It would be well in
ordering eyes from a dealer to give the measure-

ments of the required eye in hundredths of an inch. A good colored eye should not, in my opinion, have too much clear or flint glass in front of the pupil. This should be thinner and thus flatter, as seen in eyes of German manufacture. In point of perfect coloring, French eyes are the best and most expressive, but they do not have the requisite flatness and the thinness of flint which the German eyes possess. English eyes may be mentioned as third in the catalogue of quality, while America must unfortunately come last. The above remarks, however, are true only as regards colored eyes, as black eyes are almost always good, no matter where manufactured.

After the bird is placed in the required attitude, smooth the feathers with the aid of small tweezers, noting that all lines and spots are in their proper places. The primary quills should be kept in position by clamping with fine wire; that is, a piece of wire should be bent on itself like a hairpin and slipped over the edge of the wing. Spread the tail by laying it on the cross-piece of wire under it, and clamp it down with a piece of very fine wire, which is wound around each end of the cross-piece. If the tail is to be spread very widely then run a wire through the two outer quills, thus keeping them apart; though even then the clamp should be used. If a convex or concave tail is desired, bind the cross-piece in a corresponding manner. I do not, as a rule, recommend binding freshly-skinned birds, nor do I consider it necessary excepting in instances where feathers are rough. If a bird be properly mounted a few more clamps on the wings will keep it in form; then the feathers can be made to stand out as they do in nature, not lie down close to the body as if the birds were badly frightened. This is particularly noticeable with owls; a perfectly happy and contented owl, who is pursuing his vocations, has apparently a body nearly or quite twice the diameter of one that is frightened.

SECTION III.: CRESTED BIRDS. — If a bird has a crest it should be raised by gently pulling forward the skin, where it will remain in position after it is neatly arranged; but in case of a dried skin, it may be necessary to prop the crest up with a piece of cotton, moulded on the head of a pin, the point of which is sunk into the head.

SECTION IV.: MOUNTING WITH WINGS SPREAD. — In skinning for spread wings, leave in the humerus as well as the forearm, and do not detach the quills, as already mentioned. Wire the wing from the inside, and clench firmly in the body; wrap the humerus to the natural size with cotton, after fastening the supporting wire to the bone with fine wire or thread. Push both wires into the shoulders of the artificial body at once, at the same time pushing the neck-wire and body into position. This can be learned by practice. Proceed as before, but support the wings while setting on either side by long wire clamps. Be sure, however, that the supporting wire is strong enough to hold the wing in position without these, and thus when the wings are dry they will be very strong.

SECTION V.: MOUNTING BIRDS FOR SCREENS, ETC. — Proceed as in specimens with wings spread, but sometimes the wings should be cut off, sewed on on opposite sides, so that they may be reversed ; that is, the back of the wing may be toward the breast in cases where it is desired that the back of the wings and breast should show. It is usual to stretch the wings up over the head, which emerges between them. The wings had better be kept in position with strips of pasteboard fastened together with wire. Sometimes both sides of the specimen show; or, in other instances, the back is covered with paper, silk, velvet, or other material.

SECTION VI. : MOUNTING DRIED SKINS. — Soften as directed in making over dried skins, observing the caution given under that section, and have the skin very pliable. The cavities of the eyes may be filled from the mouth or from the inside of the skin. If the skin be too tender to turn, rasp it down by working through the orifice. Mount as directed in fresh specimens, but dried skins almost always require to be bound with winding cotton in order to keep the feathers in place. They also require rather more harder filling with cotton. This should be wrapped around the bird in as continuous a string as possible until all the feathers lie smoothly. They may be arranged under the bindings with small tweezers. Avoid binding too closely or too tight, and above all things bind evenly, that is, do not make depressions nor allow elevations to appear, for, as a rule, these will always remain after the bindings have been removed. Small birds should be allowed to stand at least a week in a dry place before the bindings are removed. Birds mounted from skins dry more quickly than from fresh specimens. Large birds should stand from two weeks to a month, especially if the wings be spread. To remove the binding threads, cut down the back, thus taking it all off at once.

SECTION VII. : PRICES FOR MOUNTING BIRDS. — For the convenience of amateurs, who do not always know what price to put on good work, we give our price list for mounting specimens on ornamental stands. Size from humming-bird to robin, one dollar and twenty five cents ; robin to wild pigeon, one dollar and fifty cents ; wild pigeon to grouse, two dollars; grouse, ducks, small owls, two dollars and fifty cents ; large hawks and medium-sized owls, three dollars and fifty cents ; loons and large owls, five dollars ; eagles, seven dollars. For birds with spread wings, add thirty-three and one-third per cent.

SECTION VIII. : PANEL WORK. — GAME PIECES, ETC. — Panel work is made by using only half of a specimen, the back side being turned in or removed. The specimen is mounted as usual and fastened to the picture or other design used as a back ground, by wires emerging from the side and firmly clenched in the body. Game pieces are made by simply mounting the specimen, then placing it in an attitude as if it were hanging dead. Much skill and study is required for work of this nature, for if carelessly done, it has the effect of a poor painting, but if well completed both panel and game pieces produce a pleasing effect. All such work should be usually placed behind glass, as, in fact, is true with all mounted birds, especially light-plumaged birds, which are liable to become soiled through exposure to dust. Mounted birds, not kept in moth-proof cases, should be carefully dusted at least twice a week to prevent the attacks of moths.

CHAPTER V.

MAKING STANDS.

SECTION I. : PLAIN STANDS. — The best stands for the cabinet are simple wooden ones, either of pine or other woods, turned by machinery with a simple cross-piece for perching birds. As a rule, the shaft should be about as high as the cross-piece is long, but in cases of specimens with long tails, the shaft should be somewhat higher, while the base should a little exceed in diameter the length of the perch, and should be about as thick as the shortest diameter of the other parts.

SECTION II. : ORNAMENTAL STANDS. — Papier-maché used for making ornamental stands is quite difficult to make, but following is the receipt : Reduce paper to a perfect pulp by boiling and then rubbing through a sieve. To every quart of this pulp add a pint of fine wood-ashes and a half pint of plaster. Heat this mass over the fire, and to every quart add a quarter of a pound of glue, which has been thoroughly dissolved in a glue-pot.

Mix well until it is of the consistency of putty, when it is ready for use.

In making a twig for an ordinary perch, fasten a moderately stout wire in a wooden base; wind it with cotton, larger at the base, tapering toward the end; bend it in a position and cover with a layer of papiermaché, then with a comb indicate the ridges in the bark of a tree, and add knots and excrescences as desired, by moulding small pieces with the fingers. Set aside to dry for a few days. If the papier-maché cracks it does not contain a sufficient quantity of glue, or if it shrinks too much, more ashes or plaster should be added. When dry paint with water-colors, made by adding dry paint to dissolved white glue, stirring until the mixture becomes of the consistency of cream. A quarter of a pound of glue will take up a pound of paint. Cover the bottom of the stand with this paint, or with some other color, then sprinkle profusely with smalt or mica sand. When dry, add artificial leaves to the branches by winding the stems around them. Trim the bottom of the stand with mosses and grass fastened on with glue. Stands for cases are made in a similar manner, but it is an improvement to touch the ground-work here and there with dry paint of various colors. A piece of looking-glass may be used to imitate water ; and ducks from which the lower portions have been cut away may be placed on this with a good effect. A very good stand may be made by simply winding a wire with cotton and painting the cotton. The cotton can be made into a species of papier-maché by soaking it in flour-paste. Rock work is made of either papier-maché, cork, blocks of wood, or pieces of turf painted and sanded, or by pasting stout paper over pieces of wood, and the whole structure painted and sanded. If papier-maché be used the effect may be heightened by sticking in pieces of quartz or other rock. Natural stumps, branches, etc., may be manufactured into stands or cases to advantage ; in short, with the aid of papier-maché, glue, moss, grasses, smalt, etc., nature may be imitated in a variety of ways.

CHAPTER VI.

COLLECTING MAMMALS.

MAMMALS are, as a rule, much more difficult to procure than birds, especially the smaller species. Mice occur in all localities. The white-footed mice are often found in the deserted nests of squirrels or of crows in the tree-tops. Jumping-mice are found in the meadows, under haycocks or in nests deep in the earth during winter, at which time they are in a dormant condition. Field-mice of several species occur in the meadows, where they have nests, while the house-mouse and several species of mice inhabit dwellings. All these little rodents may be trapped by using a variety of bait, and the same is true of squirrels, which are, however, quite easy to shoot. The gray, red, and flying-squirrels live in nests placed in bushes or trees or in holes in tree-trunks. Shrews and moles burrow in the ground, and they may be snared by setting fine wire nooses in their holes. Cats often bring in these little mammals and leave them lying around, as they rarely eat them. A pit dug in an open field or a barrel set down with the top on a level with the ground and half filled with water will be the means of capturing many rare, small mammals which fall into it accidentally. Mink, weasel, otter, rabbits, skunks, etc., may be trapped or shot. A variety of bait may be used to decoy animals of this class, and the contents of the scent-bags of any of these species are good ; as well as fish, birds, or small mammals. Foxes, wolves, etc., which occur in the wilder sections, may be shot or trapped, and the same is true of wild-cats, pumas, and other large mammals, in procuring which the hunter must be guided by circumstances.

CHAPTER VII.

MAKING SKINS OF MAMMALS.

SECTION I. : SKINNING SMALL MAMMALS. — Lay the animal on its back, make an incision about one-third of the length of the body on the under side of the body from the vent forward, peel down on either side until the knee-bones are exposed, then cut the joint and draw out the leg, at least as far as the heel. Remove the flesh, cover well with preservative, and turn, then proceed thus with the opposite leg. Pull down to the tail and draw out the bone by placing a stick on the under side of it and pressing backward. If the tail bone does not readily come out, as in musk-rats, wrap the tail in cloth and pound it with a wooden mallet, and it will then come out without further trouble. Peel down on either side until the front legs appear, cut off at elbow joints, and draw these out ; remove the flesh, cover with preservative, and turn. Skin over the head, taking care to cut off the ear next the skull, so as not to

86

cut through into the exterior surface; pull down
the edges, cut between the lids and eye-sockets
down to the lips, cut between these and the bone,
but near the latter, thus removing the skin entirely
from the skull; cover the skin well with preservative, after removing all fat and surplus bits of
flesh. Then turn the skin, detach the skull from
the body, by carefully cutting between the atlas,
the last vertebra joint, and the skull. The skull
should be boiled to remove all the flesh and brain ;
or, if this cannot readily be done, and if the mammal be very small, roll it in preservative, and lay it
one side ; if the animal be large, cut off all the flesh
possible, and work out the brain through the open-
ing in the base of the skull. It is always, however,
best to remove the flesh by boiling ; after which
care should be taken to tie the lower jaw firmly to
the upper.

SECTION II.: SKINNING LARGE MAMMALS. —
Large mammals should be skinned by making a
cross incision down the entire length of the breast,
between the fore-legs to the vent, then down the
under side of each leg quite to the feet. Remove
the skin but leave in two bones and the joints in
each leg. In removing the horns of a deer or other
ruminant, make cross cuts between the horns,

and then back down on the neck for a short distance.   The lips of a large mammal should be split
open carefully, and the ears turned out quite to the
tip; this can be done with a little practice.   Cover
with preservative, well rubbed in, and dry as quickly
as possible without tearing.

SECTION III.: MAKING SKINS OF MAMMALS. —
Remove all blood and dirt, by either washing or
by continuous brushing with a stiff brush.   Dry
off with preservative: rub it well into the hair.
Draw out the bones of the leg, wrap them well
with cotton to the original size of the leg; then
fill out the head to the size and form of life, sewing up the neck, and fill up to the body to the size
of nature with cotton or tow.   Sew up the orifice,
then lay the skin, belly down, with the feet laid
neatly; and if the tail is long, lay it over the back.

Mice and other small mammals should not have
the bone of the tail removed, as the skin cannot be
filled and turned over the back easily.   Large
mammals may be also made up if they are to be
used for cabinets or for skins.

SECTION IV.: MEASURING MAMMALS. — It is
quite as easy to measure mammals as birds.   The
dimensions to be taken may be seen by the accom-
panying filled blank, which is the form I always use.

Arctomys monax.

Locality.	Age.	Sex.	Date.	No.	Nose to Eye.	Nose to Ear.	Nose to Occiput.	Nose to Root of Tail.	Nose to Outstretched Hind Leg.	Tail to End of Vertebra.	Tail to End of Hair.	Hand Hind Leg.	Hand Length.	Hand Width.	Height of Ear.	Muzzle.	Girth.	Skull* Length.	Skull* Width.	Remarks.
Ipswich	Adult	♂	1868. Aug. 22	58	1.50	2.95	2.30	13.00	15.00	4.05	6.00	3.10	2.10	.78	.85	.20	—	—	—	Light colored.
"	"	♀	" 20	55	1.57	2.80	3.45	15.50	20.15	4.50	6.75	2.80	1.85	.92	.75	—	14.50	—	—	" "
"	"	♀	" 13	43	1.32	2.94	3.45	15.25	19.50	5.45	7.60	2.95	2.05	.70	.65	.15	9.75	—	—	Top of head black.

* This measurement is taken after the animal is skinned; the width of skull is measured on the widest part, the length on the longest part.

CHAPTER VIII.

MOUNTING MAMMALS.

SECTION I.: SMALL MAMMALS. — Skin as directed, but the skull should not, as a rule, be detached unless the animal be large enough to have the lips split. The eye cavities should also be filled with clay. Cut a piece of wire of the suitable size to support the head; have it about twice as long as the head and body of the specimen in hand. Wind up a turn or two with the pliers small enough to enter the cavity in the base of the skull, which will have to be enlarged to admit of the ready removal of the brains. Place the wound portion of the wire in this cavity, and fill in around it with either plaster of paris, or tamp in excelsior, tow, or cotton firmly enough to hold the skull perfectly firm on the wire. Wind up a body of excelsior or grass, as nearly the form and size of the one removed as possible, taking care that the neck be of proper shape, and that the surface be very smooth.

90

This surface may be covered with a thin layer of clay or of papier-maché, if a very nice smooth surface is required, in case of short-haired mammals. Cut four wires for the legs and one for the tail. Run the wire up the front legs, and tie them firmly to the bone with fine wire, especially at the joints. Now wind each leg with cotton, hemp, or tow to the size and form of the muscles removed. In order to get the legs very exact, one may be wound before the muscles of the other be removed, and measurements may thus be taken. The legs may be also covered with papier-maché or a thin layer of clay in short-haired mammals. Now place the body in position, taking care that the wire of the head goes the entire length of the body, and is firmly clinched.

The wires of the front legs should enter the body at the proper point on the shoulder. The wires of the hind legs should also enter the body at the point near the back, where they join the natural body. Run a wire the entire length of the tail and fasten in the lower end of the body. See that all wires are firmly clinched, and sew up the orifice. Bend the legs into as natural a position as possible, and insert the wires protruding from the soles of the feet into the holes in the

stand or perch; bend the body in position, insert
the eyes, arranging the lids carefully over them,
taking care the eye has the proper form in the
corners.

Arrange the eyelids and ears by occasionally
moulding them into form as they dry. Smooth
the tail carefully and attend to all the little details, such as spreading the toes etc., etc., and

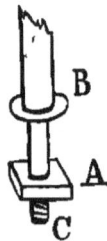

FIG. 17.

carefully watch them from day to day, until the
animal becomes perfectly dry.

SECTION II.: LARGE MAMMALS. — In drawing
the lines between mammals mounted as described
above and the present method, it may be well to
remark that the one now given is the best in all
cases, but requires rather too much time to be used
with very small specimens.   Have five large wires
or bolts of a suitable size to support the mammal
mounted, cut to the proper length, and cut a screw
on either end for about two inches (Fig. 17, *a*).

Screw a broad flat nut on (Fig. 17, *b*), then have another nut ready to screw on above the first. Prepare a strip of board a little shorter than the natural body of the mammal, and in this bore four holes, two at each end, with one extra between the two, but a little back of them on the front end. After bending the bolts so as to form the legs, place the ends in the holes and screw on the nuts, place the lower ends of the irons in the holes in the stand and screw on the nuts, thus the beginning of the structure will stand firm. Fasten the end of the fifth iron firmly in the brain cavity by filling in with plaster, or wedging in pieces of wood, and screw the lower end in place. Now wind excelsior on the legs to the proper size and form ; cover it with a thin layer of cotton. Then place on the body in sections of excelsior of exactly the form and size of life, and cover with clay. The neck is now to be formed in the same way; of course to get all the parts accurate, one must have the natural body, which has been removed, at hand, or should have the correct measurement of it. The skin, from which the leg bones have been removed quite to the toe-nails, may be fitted on occasionally to judge the effect. Procure sheet lead, and, if too thick, beat it out;

cut it in the form of the cartilage removed from the ear. Fasten wire into these pieces of lead with the ends protruding downward ; bore holes in the skull into which the ends are introduced, thus forming the support, and keeping the ears in proper position. Supply the muscles of the skull with excelsior and clay or papier-maché, then adjust the skin firmly and sew up. Fill the lips and nose with papier-maché or clay, and mould into shape. The above instructions, if followed, will give a mounted specimen, but I cannot convey the ideas which must teach the student the exact poise, the swell of the muscle, the exact shape of the eye which will give life and beauty to the subject in hand ; all these must come from patience, study, and long practice, for skilful taxidermists do not spring at once into existence, but require experience and careful education.

SECTION III.: MOUNTING DRIED SKINS OF MAMMALS. — Skins of mammals must be soaked in a strong solution of alum water, and when perfectly soft see that the parts above the lips, eyes, etc., are peeled down quite thin, and that every portion of the skin is perfectly pliable, then it should be moistened as described.

SECTION IV.: MOUNTING MAMMALS WITHOUT

ANY BONES. — If the skull of a mammal be desired
for a skeleton, a cast may be taken of the entire
head before the flesh is removed, by placing the
head in a box which will contain it and leave a
space around it ; pour in plaster of paris to the
consistency of cream, until the head is about half
covered — which should be placed on the bottom of
the box, lower jaw down — let the plaster set ; now
cover the top surface of the plaster with paint, or
oil, or paste paper over it. Then fill up the box
with fresh plaster : after this has set well remove
the side of the box and open the mould where the
joint was made with the paint or paper. Take out
the head, and then cut a hole in the mould at the
base of the skull, in which the plaster for the head
may be poured. Paint or oil the inside of the
mould everywhere, fit the pieces together, then
tie firmly and pour in the plaster for the mould ;
then insert the bolt for the head in the hole, and
let the plaster set around it. Remove the mould
by chipping off pieces with a chisel until the paint
surface is exposed. If the head be large and heavy,
a large ball of excelsior, in which the bolt is firmly
fastened, may be placed in the centre, but this
must be covered with a thin layer of clay to make it
impervious to plaster. The lips and other naked

spaces must be painted the color of life, with paint
mixed with varnish, first filling out the imperfections with paraffine wax. Casts may be taken of
the larger in wax, making a mould in plaster.

CHAPTER IX.

MOUNTING REPTILES, BATRACHIANS, AND FISHES.

MOUNTING reptiles, batrachians, and fishes as collected in this department is scarcely a part of taxidermy. I shall only give general instructions regarding mounting some species. Snakes may be readily skinned by cutting a longitudinal insertion about a fourth of the distance down from the head on the lower side where the body begins to enlarge, near its greatest diameter ; then the skin may be speedily taken off both ways. When the vent is reached the skin comes away harder, but in order to make a perfect piece of work it must be skinned quite to the end of the tail, even if it splits open ; the eyes must be removed from the inside of the head. The skin on the top of the head cannot be removed in this class of animals, leaving the jaw and skull. Cover well with preservative, and turn the skin. To mount, two ways are practised, one with plaster, in which the orifice on the inside and the vent are sewed up, and the plaster poured into the mouth until the snake is filled. It is well, however, to place a copper wire the entire length of the animal to strengthen it ; then before the plaster is set, place the snake in the proper attitude. This kind of work requires practice, as you must be careful of the attitude in which you wish to place the animal, as the plaster begins to set quite quickly ; to make it set more slowly, however, mix in a little salt. The mouth should be filled up with clay or plaster. Care should be taken that water does not accumulate in any portion of the skin, and it should be perforated with an awl occasionally to allow the water to escape. The skin of a snake may be filled with papier-maché by working small pieces downward ; then insert a wire and place into position. The skin requires some time to dry, and in both cases place the mounted reptile in a dry place, where it will rapidly dry, as the skin is liable to decay if kept in a damp place.

SECTION I. : MOUNTING LIZARDS, ALLIGATORS, ETC. — Reptiles of this description should be skinned like mammals, through a longitudinal insertion made in the abdomen. The skin from the top of the head cannot be removed however. In mounting, proceed exactly as in mammals, but as there is no hair to hide defects, all cotton, excelsior, etc., wound on the bones must be very smooth. The attitudes of all this class of animals are apt to be stiff and ungainly even in life; but by putting a bend or two into the tail, turning the head, or slightly curving the body, too much rigidness may be avoided.

SECTION II.: MOUNTING TURTLES. — To remove the skin from a turtle, cut away a square portion of the under shell, using a small saw for this purpose. Then remove the softer portion through this hole, and draw out the legs and head as in mammals; but the top of the head cannot be skinned over. In mounting proceed as nearly as possible as in mammals, only the legs may be filled with clay or plaster in small specimens. Care should be taken not to fill the skin too full; but let the wrinkles show, as seen in life, and imitated as nearly as possible.

The shell of the soft-shelled turtle, like the leather-back, is quite difficult to keep in good condition — is apt to become distorted in drying. The only method which has occurred to me is to cover the body, and exposed under portions, with layers of plaster, which will keep the shell in position until it is dry, when it may be removed.

SECTION III. : MOUNTING FISHES. — Fishes are

quite difficult to skin, especially those with scales. In flat fishes I remove a portion of one side, skinning the other; then, in mounting, lay the animal on its side. Mounting in this case means filling the fish to its natural life-size with cotton, tow, or other available material. Plaster or clay will also answer. The fins may be pinned out flat against pasteboard, or put in place with fine wire.

In skinning larger fishes, or those which have no scales, or scaled fishes which have cylindrical shaped bodies, open from beneath by cutting nearly the whole length of the body. The skin from some fishes comes off easily, while in others it is more difficult to remove. In mounting large fishes use a hard core to the body, made of either wire or wood. The fins should be wired from the inside; care should be taken that the skin lies smoothly over the surface beneath, as it shows considerably in drying, and all imperfections around it.

In preserving the skins of all reptiles and fishes the dermal will be found excellent, especially in removing the oil from the skins, etc. Cover well with the preservative, and nothing more will be necessary. Skins of this class of animals may be kept for future mounting by simply coating with the preservative, and kept turned wrong side out without filling. When they are to be mounted throw them into water, in which a small quantity of dermal has been dissolved. When they are soft turn and mount as in fresh skins.

INDEX.

A.

D.

P.

Q.

R.

S.